The FCC Rule Book

Complete Guide to the

FCC Regulations

Governing Amateur Radio

Edited by
Norman Bliss, WA1CCQ

Published by
The American Radio Relay League

Contents

Foreword

Some exciting things have happened since the previous edition of *The FCC Rule Book* appeared. Limited access to 219-220 MHz is now available for digital links. The FCC approved a vanity call sign program allowing amateurs to get call signs of their choice. Clubs can once again obtain new station licenses. There have been several changes affecting digital modes, including two-way automatic digital HF operation in specified subbands and one-way operation elsewhere under certain circumstances. The FCC implemented electronic filing from VECs, dramatically reducing the turnaround time for new licenses and upgrades from months to days, and made it possible to use the license as soon as it is granted. The FCC itself has been reorganized.

The tenth edition of *The FCC Rule Book* reflects these changes. It too has been reorganized, and contains the most current information available at press time. The appendices have been expanded to include more original material.

Amateur Radio is defined by its own set of Federal regulations, and amateurs need to stay up-to-date with them. In this book you will find a description of the FCC rule-making process, and how *you* can participate. And you'll find such useful reference information as the FCC's call sign assignment system, international regulations, a listing of countries permitting third-party traffic handling with US amateurs, operating abroad, band plans, testing procedures and much more.

The FCC regulations are designed to allow great flexibility in developing your operating skills and technical knowledge. *The FCC Rule Book* is designed to be consistent with this philosophy. We hope it will help you get the most out of your operating time.

David Sumner, K1ZZ
Executive Vice President
Newington, Connecticut
August 1995

Acknowledgments

The FCC Rule Book is the product of many individuals' expertise and hard work. Dale Clift, NA1L, and Bob Halprin, K1XA, who developed the local antenna regulation discussion, were major contributors.

Bruce Hale, KB1MW, contributed material on specialized operating practices, technical standards and repeaters. Edward Mitchell, KF7VY, contributed substantially on the regulatory issues surrounding emergency communications. ARRL Laboratory Supervisor Ed Hare, KA1CV, contributed much of the material in the EMI/RFI chapter. Technical Relations Manager Paul Rinaldo, W4RI, is responsible for most of the material on digital communications and emission standards.

I wish to particularly thank Rick Palm, K1CE, for his assistance with the present edition and for the strong foundation on which it was built. I also wish to thank Regulatory Information Specialist and "Washington Mailbox" columnist John Hennessee, KJ4KB, for his invaluable assistance throughout this project, and Book Team Supervisor Joel Kleinman, N1BKE, who guided this book through the publication process and helped put the polish on. Thanks also go to Zack Lau, KH6CP, and Lisa Kustosik, KA1UFZ, for their help. Overall production work and page layout was done by Dan Wolfgang. Dave Pingree, N1NAS, handled graphics, and Sue Fagan designed the cover.

Norman Bliss, WA1CCQ
Editor

What's New

- The FCC approved the vanity call sign program and began issuing new club station licenses. See Chapter 2 and §§97.3, 97.17, 97.19 and 97.21.

- The FCC dramatically sped up amateur licensing by allowing VECs to file examination data electronically, and permitting new amateurs to operate as soon as their licenses are entered into the FCC database. See Chapter 2.

- The Technician Plus is now a separate license class. Holders have passed the Technician theory and a 5 wpm code test, and have limited HF privileges. All previous Technicians who have passed a code test are now Tech Plus. The Technician class now consists solely of persons who have not passed a code test; they have no HF privileges.

- 219-220 MHz was allocated to amateurs on a secondary basis for digital message forwarding networks only. See Chapter 5 and §§97.201, 97.301, 97.303, 97.305, 97.307, and 97.313.

- §2.106 and Part 15 of the rules were modified to give amateurs primary status in the 2390-2400 MHz and 2402-2417 MHz bands. See Chapter 5.

- The FCC approved automatic HF digital operation in specific subbands. See Chapter 6 and §§97.109 and 97.221.

- Congress passed Public Law 103-408 declaring that reasonable accommodation should be made for amateur operation from residences, private vehicles and public areas. It doesn't require compliance and cannot be used in court, but does strengthen amateur cases before municipal bodies and homeowner's associations. See Appendix 6.

- The FCC underwent a major reorganization at the end of 1994. See Chapter 15.

- The FCC is now on the Internet. See Appendix 12.

- The FCC Forms 610 and 610-V are now available via fax from the FCC's Fax-On-Demand service. White paper copies are acceptable, but you must photocopy thermal paper output. See Appendix 12.

- The FCC changed its rules to clarify that amateurs are permitted to use the PacTOR, G-TOR and CLOVER digital modes. See §97.309.

- The FCC has begun sending out a renewal short form, the 610-R. It's a personalized form, and all you need to do is sign and return it. It is not available as a blank form.

At Press Time

- As of February 1996, the FCC had not begun the vanity call sign program; release of Form 610-V was delayed pending review of several petitions for reconsideration. The fee will be $30 for a 10-year license.
- The FCC issued a Notice of Proposed Rule Making (WT Docket 95-57) proposing five rule changes. The first would designate one member of a VE team as examination session manager. The second would allow former licensees to get examination credit for any exam previously passed. A third proposal would increase the number of members required for a club to be eligible for a station license from two to four. Fourth are proposed rules for the issuance of 1×1 special event call signs. Finally, the FCC proposes to relax placement of portable indicators. For more information, see page 82 of July 1995 *QST*.
- A second NPRM in ET Docket 94-32 invited comments on possible new restrictions on amateur operations in the 2390-2400 and 2402-2417 MHz bands, despite our upgrade to primary status there.
- Among other proposals in ET Docket 94-124, the FCC has proposed opening 76-77 GHz for automotive radar systems and to "temporarily" disallow amateur use of this portion of the 75.5-81 GHz band. To balance this loss, the FCC would upgrade amateur status at 77.5-78 GHz to coprimary. We are currently primary in the 75.5-76 GHz portion and secondary in the 76-81 GHz portion.
- The ARRL has petitioned the FCC to implement US participation in an *International Amateur Radio Permit* (IARP) covering countries in the Western Hemisphere. The US signed an Inter-American Convention establishing the IARP; now the FCC must issue rules to implement it. The permit would allow holders to operate temporarily in other countries which have signed the convention.
- Congress passed and the President signed into law sweeping reforms in the Telecommunications Act of 1996. The only item in the bill affecting amateurs is elimination of "unnecessary conflict-of-interest provisions for publishers and equipment manufacturers" and of annual financial certification requirements from VEs. These changes are reflected in Appendix 1, and the FCC has changed its rules accordingly. See §§97.509, 97.521, and 97.527.

The Amateur's Code

The Radio Amateur is:

CONSIDERATE...never knowingly operates in such a way as to lessen the pleasure of others.

LOYAL...offers loyalty, encouragement and support to other amateurs, local clubs, and the American Radio Relay League, through which Amateur Radio in the United States is represented nationally and internationally.

PROGRESSIVE...with knowledge abreast of science, a well-built and efficient station and operation above reproach.

FRIENDLY...slow and patient operating when requested; friendly advice and counsel to the beginner; kindly assistance, cooperation and consideration for the interests of others. These are the hallmarks of the amateur spirit.

BALANCED...radio is an avocation, never interfering with duties owed to family, job, school or community.

PATRIOTIC...station and skill always ready for service to country and community.

—The original Amateur's Code was written by Paul M. Segal, W9EEA, in 1928.

CHAPTER 1

Introduction

A mateur Radio means different things to different people. That's one of its beauties—diversity. Like a fine-cut gem, it shines in many directions. Steeped in tradition, Amateur Radio means to old-timers the memory of a past era of racks of black, wrinkle-finished, meter-spattered radio chassis replete with glowing tubes and humming transformers. To many newer amateurs, it equates to "high-tech" and the opportunity to pioneer and develop the latest experimental digital code, data network or satellite technique. To others, it means things perhaps less heady—simply an opportunity to make friends around the world on the long-haul high-frequency bands or across town through the local repeater. It means community service—the chance to use radio to help fellow human beings in need, whether by assisting safety services deal with a tornado watch or by organizing a local walkathon. To students, it means a head start on a career path. To most, it's a little of all of the above.

A common thread in Amateur Radio's history has been a dynamic regulatory environment that has nurtured technological growth and the diversity described above. This thread continues to sew together the elements of Amateur Radio today and prepare it for tomorrow's challenges.

This chapter looks at the basics of the present regulatory structure as a launching pad for discussions in the rest of the book. There's no better place to start than with a basic definition of Amateur Radio.

Amateur Radio Defined

The Federal Communications Commission (FCC) has a good definition for Amateur Radio. It's near the beginning of the Amateur Radio rules, officially known as Part 97. No one has yet come up with a better basic definition:

A radiocommunication service for the purpose of self-training, intercommunication and technical investigations carried out by amateurs, that is, duly authorized persons interested in radio technique solely with a personal aim and without pecuniary interest [97.3(a)(4)].

Note that phrase "without pecuniary interest." The service is "amateur" because it's strictly noncommercial; you may not earn money or receive any other material compensation for your transmissions. Amateur Radio is for people interested in the technical and communications aspects of radio and those who are able to provide emergency communications in disasters, all for the benefit of the general public.

What is the FCC?

The Federal Communications Commission (FCC) is the US government agency charged by Congress with regulating communications involving radio, television, wire, cable and satellites. This includes Amateur Radio. The objective of the FCC is to provide for orderly development and operation of telecommunications services.

The FCC functions like no other Federal agency. It was created by Congress and it reports directly to Congress. The FCC allocates bands of frequencies to nongovernment communications services and assigns operator privileges (the National Telecommunications and Information Administration allocates government frequencies).

Amateur Radio began with a few experimenters in the early 1900s and has grown to nearly 700,000 licensed operators in the US alone. Amateur Radio represents principles of radio communications that have endured and advanced since the days of the earliest radio pioneers.

Basis and Purpose

The FCC expands on its basic definition in Section 97.1 of the rules, which sets forth Amateur Radio's "Basis and Purpose." Let's have a look at each of the five basic principles:

Recognition and enhancement of the value of the amateur service to the public as a voluntary noncommercial communication service, particularly with respect to providing emergency communications [97.1(a)].

Probably the best-known aspect of Amateur Radio in the public eye is its ability to provide lifesaving emergency communications when normal means of contact are down. In hurricanes, earthquakes, tornadoes, airplane crashes, missing-person cases, and other accidents and disasters affecting the civilian population, Amateur Radio often provides the first means of contact with the outside world—a communications "first responder." Red Cross and civil preparedness agencies rely heavily on the services of volunteer radio amateurs.

One of the most important aspects of the service is its noncommercial nature. Amateurs are prohibited from receiving any form of payment for operating their stations. This means that hams, whether assisting a search-and-rescue operation in the high Sierra, relaying health-and-welfare messages from a hurricane-ravaged Caribbean island or providing communications assistance at the New York City Marathon, make their services available free of charge. Amateurs operate their stations for the benefit of the public and for their own personal enjoyment only.

Continuation and extension of the amateur's proven ability to contribute to the advancement of the radio art [97.1(b)].

For more than 80 years, hams have carried on a tradition of learning by doing, and since the beginning have remained at the forefront of technology. Through experimentation and building, hams have pioneered advances, such as techniques for single-sideband transmissions, and are currently engaged in state-of-the-art designs in digital radio and spread-spectrum techniques. Amateurs were among the first to bounce signals off the moon to extend signal range. Hams' practical expe-

STATE OF FLORIDA

Office of the Governor

THE CAPITOL
TALLAHASSEE, FLORIDA 32399-0001

LAWTON CHILES
GOVERNOR

October 13, 1992

Mr. David Sumner
Executive Vice President
The American Radio Relay League
225 Main Street
Newington, Connecticut 06111

Dear Mr. Sumner:

On behalf of the state of Florida, I am writing to thank the many amateur radio operators who assisted in the Hurricane Andrew relief effort.

Scores of amateur radio operators rallied to South Florida from across the United States, helping to provide desperately needed communications to local, state, and federal agencies.

They provided moral and physical support to local amateur radio operators, many of whom had suffered severe damage to their homes, yet provided around-the-clock communications at emergency operation centers, food distribution centers, and field medical facilities.

Hundreds more assisted at their home stations around the country, passing health and welfare messages to concerned relatives of south Floridians.

The amateur radio service can be proud of its members, who time and time again serve the country unselfishly. The state of Florida owes them a debt of gratitude and thanks.

With kind regards, I am

Sincerely,

LAWTON CHILES

Florida Governor Lawton Chiles sent this letter to the ARRL expressing thanks on behalf of his state for the communications assistance rendered by amateurs following the devastation of Hurricane Andrew in 1992.

rience has led to technical refinements and cost reductions beneficial to the commercial radio industry.

Encouragement and improvement of the amateur service through rules which provide for advancing skills in both the communication and technical phases of the art [97.1(c)].

This obligates the Commission to see that its rules allow room for amateurs to move in new directions so they can live up to their potential for advancing technical and communications skills. The amateur service is constantly changing to meet new challenges; a flexible and dynamic rule structure provides for this. As you will see in Chapter 15, amateurs play an important role in the rulemaking process.

Expansion of the existing reservoir within the amateur radio service of trained operators, technicians and electronics experts [97.1(d)].

The greater the number of amateurs proficient in communications techniques and electronics, the greater the resource Amateur Radio is to the public. For example, during WWII, hams were a significant resource to the military—many were called upon to be radio operators and technical experts, and many hams fought and died for the interests of our country.

Continuation and extension of the amateur's unique ability to enhance international goodwill [97.1(e)].

Hams are unique—they can travel to the far reaches of the Earth and talk with other amateurs in foreign countries simply by stepping into their radio "shacks." H. G. Wells had his time machine—hams have space machines!

In this time when global community initiatives are so important, amateurs can make substantial contributions by representing their respective countries as ambassadors of goodwill. Amateur-to-amateur communications transcend cultural boundaries between societies. Amateur Radio is a teacher in Lincoln, Nebraska, trading stories with a London boarding school headmaster, or a tropical fish enthusiast in Alabama learning about different species in the Amazon River from a missionary stationed in Brazil; it's a means of making friends everywhere.

Purpose of the Rules

It's the Commission's responsibility to see that amateurs are able to operate their stations in a manner consistent with the objectives outlined above. The FCC must also ensure that hams have the knowledge and ability to operate powerful and potentially dangerous equipment safely without causing interference to other radio services.

To these ends, the FCC requires examination of all amateurs in theory, operating, regulations and, except for the Technician class license, Morse code before a license is issued. The Commission fashions its rules in accordance with the Basis and Purpose of Amateur Radio. The rules must be written to create a flexible regulatory environment that encourages, rather than limits, development of Amateur Radio and its ability to use modern technology. The FCC reviews its rules from time to time to remove uncertainty, incorporate existing interpretive policy where necessary, and delete unnecessary or obsolete rules.

Amateur Radio is akin to our public parks: it requires protection from exploitation or conversion to commercial use so the public can use it and benefit from its use. The rules are designed to help protect Amateur Radio from such encroachment.

Basis for the Rules

The basis for the FCC's regulations is found in treaties, international agreements and statutes that provide for the allocation of frequencies and place conditions on how the frequencies are to be used. For example, Article 32 of the international Radio Regulations limits the types of international communications amateur stations may transmit, and mandates that the technical qualifications of amateur operators be verified.

The Commission's statutory authority to make operational and technical rules, authorize frequencies and license amateurs and stations in the US comes from the Communications Act of 1934. For a comprehensive discussion of the Communications Act of 1934, as well as the International Telecommunication Union regulations, refer to Chapter 15.

Three other acts of Congress affect FCC rulemaking. The Administrative Procedure Act allows the public to participate in the rule making process. You may petition the FCC for rule changes, comment on current proposals, and ask the FCC to reconsider decisions. The FCC must take into consideration your input when making rules. There's more on the Administrative Procedure Act in Chapter 15.

Under the Regulatory Flexibility Act, the Commission certifies that new rules will not have a significant economic impact on a substantial number of small entities, such as businesses.

Finally, the FCC analyzes its rules under the Paperwork Reduction Act to make sure they tend to decrease the information collection burden imposed on the public.

A Quick Journey through Part 97

The Amateur Radio service rules, Part 97, are organized in six major subparts: **General Provisions**, **Station Operation Standards**, **Special Operations**, **Technical Standards**, **Providing Emergency Communications** and **Qualifying Examination Systems**. Here's a brief description of the purpose and discussion of the highlights of each subpart:

General Provisions

Subpart A covers, as its name suggests, basics that apply to all facets of Amateur Radio. The "Basis and Purpose" of Amateur Radio, discussed above, is found at the beginning [97.1]. Established in 1951, these principles have withstood the test of time.

Definitions of key terms used throughout the body of the rules are presented here and form the foundation of Part 97. They are the keys to unlocking its mysteries. Before heading into the rest of Part 97, make sure you're familiar with these definitions [97.3].

Subpart A contains a useful incorporation of FCC policy on limited preemption of state and local restrictions on amateur antenna installations. The policy has encouraged open cooperation and dialogue between the communities seeking to regulate ham towers and antennas, and the amateur community itself. Almost without exception, communities that have sought to regulate amateur stations unreasonably have done so without advance knowledge of the FCC's policy. The codification of the policies in the rules should help bring them to communities' attention [97.15(e)]. For more information about antennas and local regulations, see Chapter 9.

The remainder of the subpart is devoted to other issues involving licensing and station location, such as authorization, antenna location and restrictions, control operators, license classes and call signs, and stations on boats and airplanes.

Station Operation Standards

Subpart B, "Station Operation Standards," covers basic operating practices that apply to all types of operation. Amateurs must operate their stations in accordance with good engineering practice and must share the frequencies with others—no one ham or group has any special claim to any frequency [97.101(a) and (b)].

The question of who is responsible for the operation of a station is covered in this subpart: The station licensee is always responsible, except where the control operator of his station is someone else, in which case both share responsibility equally [97.103(a)].

The requirements of control operators and station control are also addressed in Subpart B. Each station must have a control point where operation of the transmitter is effected [97.109(a)]. A control operator must always be at the control point, except in a few cases where the transmitter is controlled automatically [97.109(b),(c),(d), and (e)].

At the heart of Amateur Radio is the principle of *two-way* communication [97.111(a)]. There is no room for one-way operation, except in a few cases: You can send a one-way transmission to make adjustments to your equipment (tuning up, for example), call CQ, remotely control devices, communicate information in emergencies, and send code practice and information bulletins of interest to amateurs [97.111(b)]. Broadcasting to the public is strictly prohibited [97.113(b)].

A major portion of this subpart, "Prohibited transmissions," covers many "no-no's," such as getting paid for operating your station, conducting communications on behalf of your employer, music, obscenity, news gathering, false signals and ciphers [97.113]. Because of the serious nature of these prohibitions and because some are open to considerable interpretation, authorized and prohibited transmissions are discussed in detail in a separate chapter.

Rules covering what an amateur can and cannot do when approached by the news media for information are also addressed in this subpart. Amateur Radio has been exploited by the broadcast media in its efforts to cover major disasters. The rules protect Amateur Radio from overzealous journalists who would use hams as alternatives to more appropriate—and expensive—communications services [97.113(b)].

Station identification, an important aspect of any amateur communication, is addressed in this subpart. The purpose of station identification is to make the source of transmissions known to those receiving them, including FCC monitors. The rules cover ID requirements for the various operating modes [97.119].

Third-party traffic rules are covered in this subpart. These specify the restrictions on stations sending messages to foreign countries on behalf of other people [97.115]. "Third-party communications" are defined in Subpart A [97.3(a)(44)]. The rules were clarified in 1989 to permit otherwise-prohibited third-party traffic in cases where the third party is an amateur licensee. This is consistent with the policies of virtually all nations' telecommunications administrations [97.115(a)(2)]. VHF and UHF message forwarding stations may be automatically controlled while transmitting third-party communications when part of a message forwarding system [97.109(e)].

A final section addresses restricted operation and sets forth the conditions that must exist in an interference case involving a neighbor's TV or radio before the Commission can impose "quiet hours," hours of the day when an amateur may not operate his or her transmitter [97.121(a)].

Special Operations

Subpart C, "Special Operations," addresses specialized activities of Amateur Radio: Auxiliary links, beacon operation, repeater operation, data and RTTY communications, telecommand (remote control) functions, satellite operations and message forwarding systems.

Repeater operation is part of many amateurs' daily activity. The rules for repeaters and auxiliary links should be understood by every trustee and user. An important regulatory approach to solving interference problems between repeaters is addressed here: Repeater station licensees are equally responsible for resolving a problem, unless one of the repeaters is approved for operation by a recognized repeater coordinator for the area and the other is not. In this case, the uncoordinated repeater has primary responsibility to resolve the problem. The station might even have to stop operation [97.205(c)]. Repeater control operators are not responsible for violative communications if the retransmission was inadvertent [97.205(g)]. Also addressed are ancillary versus control functions, and the right of repeater owners to limit access [97.205(e)].

Technical Standards

The word *standard* implies consistency and order—and this is what Subpart D, "Technical Standards," is all about. The Commission has made these standards a basic framework so all kinds of amateur operation can live peacefully in the same house, and so the Amateur Radio household as a whole can live peacefully with other radio occupants in the spectrum neighborhood.

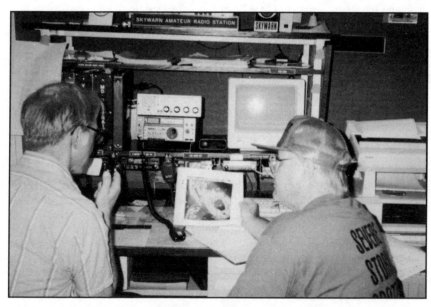

Jan Jubon, K2HJ (l), and Joe Peters, WB4WZZ, cover the night shift at the SKYWARN Amateur Radio station in the Washington, DC, forecast office of the National Weather Service. The time is early Monday morning, August 24, 1992, and Peters holds the latest printout of a GOES satellite image showing Hurricane Andrew as it begins to touch land in Florida. SKYWARN in Washington was activated to provide a backup radio link, if needed, between the National Hurricane Center in Coral Gables, Florida, and the backup hurricane center at the National Meteorological Center in Camp Springs, Maryland. The amateur HF transceiver was used to monitor hurricane nets and 2 meters was used to coordinate local amateur staffing for the SKYWARN backup station. *(photo courtesy of SKYWARN Net Manager Dan Gropper, KC4OCG).*

Amateurs may operate their stations in certain segments of the radio spectrum specified in this subpart [97.301]. In addition to further restrictions by license class [97.301(a) through (f)], there are limits on the different modes (emissions) of operation [97.305, 97.307, 97.309 and 97.311].

In many cases, amateurs share operating privileges on a primary or secondary basis with other radio services. Depending on the status of the amateur occupancy of a particular band, amateurs may have to avoid causing interference to, and tolerate interference from, other services [97.303].

Amateurs must use the minimum amount of power necessary to carry out the desired communications effectively and reliably. An amateur doesn't need a planet-crushing amplifier to reach his friend across town on 2 meters! Some amateurs forget or disregard this rule in their zeal to have the biggest signal on the band, a condition once described by Dr Sigmund Freud [97.313(a)].

Providing Emergency Communications

Subpart E, "Providing Emergency Communications," is brief, but important. It enjoys a prominent place in the rules and addresses disaster communications, stations in distress, communications for the safety of life and protection of property, and the Radio Amateur Civil Emergency Service (RACES).

When it comes to supplying emergency communications, e.g., communications related to the *immediate* safety of life, and/or the *immediate* protection of property, in the absence of other communications facilities, the rules permit amateurs wide latitude [97.403]. The rules permit the FCC to specify conditions and rules for amateur operation during an emergency situation. The Commission can set aside specific frequencies for emergency use only [97.401(c)].

RACES uses amateurs for civil defense communications during local, regional or national civil emergencies. To participate, you must be enrolled as a member of a local civil defense agency [97.407].

Qualifying Examination Systems

The final subpart of the rules, Subpart F, deals with the examination system. It covers such things as qualifying for a license, exam requirements, elements and standards. Test procedures are specified and exam preparation is covered.

In December 1983, the Commission delegated much of the exam administration program to amateurs themselves. The rules provide a system of checks and balances on volunteer examiners (VEs), who administer exams at the local and regional levels. This protects against fraud and provides integrity for the exam process.

Rule Interpretations

The purpose of this book is to provide you with a complete set of the rules and a commentary on them to help amateurs apply the rules in specific situations. The following chapters provide time-tested and FCC-approved interpretations, and copies of FCC letters and news releases describing the application of the rules. Graphic depictions of rules are provided where possible to lend clarity—maps, portrayals of antenna structures near airports and glide-slope rules are examples.

There are other parts of the Commission's rules that affect Amateur Radio. In many cases, they are referenced in Part 97. This book discusses these rules and provides the text for reference.

A Final Note

Generally, the rules are meant to be inclusive, rather than exclusive. Some hams stay up all night thinking of ways that a certain interpretation will prevent them from doing something with their radio. *Operating flexibility* is the name of the game—the rules provide a framework within which amateurs have wide latitude to do all kinds of constructive things in accordance with the basis and purpose of the service. The rules should be viewed as positive vehicles to promote healthy activity and growth, not as negative constraints that lead to stagnation.

CHAPTER 2
Your License, Passport to the World

Under international agreements, the FCC is obligated to ensure that you have a license before you operate on amateur frequencies [international Radio Regulations (RR) Article 32, Sec. 3(2)] and that you can operate your station safely while limiting interference to others [RR, Article 32, Secs. 3(2); 4; and 5]. It must also make sure you can copy Morse code if you want to operate on the HF bands [RR, Article 32, Sec. 3(1)]. Thus, the Morse code examination for FCC licenses that permit HF operation is an international requirement. And finally, the FCC does what it can to encourage you to develop and improve your operating and technical skills [97.1(c)].

To meet these ends, the Commission manages a time-tested licensing program consisting of six grades of license—two entry-level classes and four others—each dangling a bigger carrot of increased operating privileges. To earn additional privileges, you must pass tests demonstrating increased proficiency in Morse code, and knowledge of the rules, operating procedures and technical theory.

Get a License First

If you are controlling a transmitter on any frequency where Amateur Radio operation is regulated by the FCC, you must have authorization to do so from the Commission. If you don't, you will be in serious trouble. Every operator must have this authorization; the FCC makes no exceptions. In most cases, you will be granted an operator/primary station license, FCC Form 660 [97.5]. Anyone is eligible to apply for an amateur license except a representative of a foreign government [97.5(b)(1)]. You don't have to be a US citizen to hold a US amateur license.

A person's amateur license is actually two licenses in one: The *operator* license specifies your license class and privileges, and allows you to be a control operator of a station [97.9(a)]. The *primary station* license permits operation of all transmitter equipment under the physical control of the licensee at locations where Amateur Radio is regulated by the FCC [97.5(a)]. The license form lists your mailing address and call sign of your station. The FCC requires that you notify them when your mailing address changes by filing for a modification of your license on Form 610. The FCC can suspend or revoke your license if their mail to you is returned as undeliverable [97.23].

Additional types of station licenses and permits exist for clubs [97.5(b)(2)], military recreation facilities [97.5(b)(3)], civil defense centers (RACES stations)

[97.5(b)(4)] and foreign amateurs visiting the US [97.5(c)(1)]. The FCC no longer issues *new* RACES station licenses, but it will renew and modify existing ones. Effective March 24, 1995, the FCC began once again to issue new club and military recreation station recreation licenses.

Because the FCC now considers its database to be the final authority for its license grants, it no longer requires that the original written authorization or a photocopy be retained at the station—but it's still a good idea! [FCC Report and Order, October 17, 1994.]

Examination is always required for a new amateur license and for each change in license class. The FCC makes no exceptions [97.501].

Classes of License

Becoming a ham is pretty easy. With a little studying, nearly anyone can do it. There are two entry-level licenses—take your choice! Table 1 shows the operating privileges for each class of license.

Like a driver's learner permit, the *Novice* license affords enough basic privileges to give the operator experience and understanding before he or she enters the mainstream. This entry level license allows you to use voice, Morse code and other modes, depending on the band, and gives ample opportunities for local, regional and worldwide contacts.

A candidate must pass a 5-word-per-minute (WPM) Morse code test [Element 1(A)], and a written exam in elementary theory and regulations [Element 2] [97.501(f)]. The written exam has 30 questions concerning the privileges of the Novice license; 22 must be answered correctly to pass [97.503(b)(1)].

The second type of entry-level amateur license is the *Technician*. It gives the holder all amateur privileges above 30 MHz. The attraction here is that, since February 1991, *there is no Morse code requirement*. Candidates for Technician must pass written examinations for Elements 2 and 3(A) [97.501(e)]. There are 25 questions on the Element 3(A) exam, 19 of which must be answered correctly to pass [97.503(b)(2)].

The next class of license is *Technician Plus*, which was created as a separate class on December 20, 1994. A candidate qualifies for this class by passing the 5 WPM code exam [Element 1(A)] in addition to the Technician class written Elements 2 and 3(A) [97.501(d)]. In addition to the Technician's privileges above 30 MHz, a Technician Plus is granted current Novice HF privileges. All Technicians licensed before December 20, 1994 and who have passed a code test are now Technician Plus licensees.

Next is the *General* class license, considered the ticket to mainstream Amateur Radio. The license conveys most HF privileges where you can participate in mainstream activities such as DXing, net operation, RTTY and contesting. Because of the extensive privileges on the "low bands," the written exam [Element 3(B)] focuses on the regulatory, operating and technical aspects of HF operation [97.501(c)]. The exam has 25 questions concerning these additional privileges, 19 of which must be answered correctly to pass [97.503(b)(3)]. Additionally, the applicant must pass a more difficult (13 WPM) code test [Element 1(B)], [97.501(c)].

If you can demonstrate technical knowledge at the intermediate level [Element 4(A)], the *Advanced* class license offers additional HF phone privileges [97.501(b)]. The exam contains 50 questions dealing with the additional privileges of the Advanced class, 37 of which must be answered correctly to pass [97.503(b)(4)]. The code requirement is the same as for the General class license—

Table 1
Amateur Operator Licenses†

Class	Code Test	Written Examination	Privileges
Novice	5 WPM (Element 1A)	Novice theory and regulations (Element 2)	Telegraphy on 3675-3725, 7100-7150◊ and 21,100-21,200 kHz with 200 W PEP output maximum; telegraphy, RTTY and data on 28.100-28,300 kHz and telegraphy and SSB voice on 28,300-28,500 kHz with 200 W PEP max; all amateur modes authorized on 222-225 MHz, 25 W PEP max; all amateur modes authorized on 1270-1295 MHz, 5 W PEP max.
Technician	None	Novice theory and regulations; Technician-level theory and regulations. (Elements 2, 3A)*	All amateur privileges 50.0 MHz and above.
Technician Plus	5 WPM (Element 1A)	Novice theory and regulations; Technician-level theory and regulations. (Elements 2, 3A)*	All Novice HF privileges in addition to all Technician privileges.
General	13 WPM (Element 1B)	Novice theory and regulations; Technician and General theory and regulations. (Elements 2, 3A and 3B)	All amateur privileges except those reserved for Advanced and Amateur Extra class.
Advanced	13 WPM (Element 1B)	All lower exam elements, plus Advanced theory. (Elements 2, 3A, 3B and 4A)	All amateur privileges except those reserved for Amateur Extra class.
Amateur Extra	20 WPM (Element 1C)	All lower exam elements plus Extra-class theory. (Elements 2, 3A, 3B, 4A and 4B)	All amateur privileges.

†A licensed radio amateur will be required to pass only those elements that are not included in the examination for the amateur license currently held.

*If you hold a valid Technician class license issued before March 21,1987, you also have credit for Element 3B. You must be able to prove your Technician license was issued before March 21,1987 to claim this credit.

◊Varies according to region; see Chapter 5.

13 WPM [Element 1(B)], [97.501(b)].

At the top is the *Amateur Extra* class license, conveying all amateur privileges, and, accordingly, the right to operate on segments reserved exclusively for Extras, away from the crowds on the rest of the band. As expected, the written examination [Element 4(B)] is the most challenging [97.501(a)]. It contains 40 questions on advanced techniques, 30 of which must be answered correctly to pass [97.503(b)(5)]. The applicant must demonstrate proficiency in Morse code at 20 WPM [Element 1(C)], [97.501(a)].

A list of the number of questions required for each specific topic on each exam is contained in the Rules [97.503(c)].

Exam Credit

When you upgrade, you will be required to pass only the test elements for the new license that are not included in the exam for the amateur license you currently hold [97.505(a)(1)]. For example, when you upgrade from General to Advanced, you will be required to pass only Element 4(A); you don't have to take Element 3(B) again. Your license must be unexpired (or expired less than two years) to qualify for credit.

Credit for any element is given to applicants who produce a valid *Certificate of Successful Completion of Examination* (CSCE). For example, if you take the Amateur Extra class exam, pass the code, but fail the written test, you will receive a CSCE for the code element. Because CSCEs are valid for 365 days, you've got a year to use it for credit at other sittings for the Amateur Extra class exam [97.505(a)(6)].

If you're trying for Technician or higher and have just passed the Novice, but don't have the ticket in hand yet, you can get credit for the Novice elements [Elements 1(A) and 2] by showing the examiners a CSCE indicating you qualified within the previous 365 days [97.505(a)(6)].

Anyone holding a current (or expired less than five years) *commercial radiotelegraph* operator license or permit may receive credit for *any* amateur code test [97.505(a)(7)].

An applicant holding a Technician license issued before March 21, 1987 (the date when new rules provided for separate General and Technician written tests) receives credit for the General written exam [Element 3(B)] [97.505(a)(8)].

No other exam credit is allowed on the basis of holding or having held any other license [97.505(b)].

Code Credit for Handicapped

The 13-WPM code test [Element 1(B)] and the 20-WPM code test [Element 1(C)], but not the 5-WPM code test [Element 1(A)], may be waived in cases where a physician has certified an applicant's inability to pass the exam because of a physical handicap. The physician must complete the *Physician's Certification of Disability* on the back of Form 610, which certifies that because of a severe handicap, the candidate is unable to pass a 13 or 20-WPM Morse code test. To be considered a "severe handicap," the disability must extend for more than 365 days after the date of the physician's certification [97.505(a)(10)]. The FCC has also stated that the individual's handicap must be so severe that the passing of the upgrade code is not possible, even with special accommodative procedures [Report and Order, Docket 90-356, August 29, 1991]. Special accommodative procedures are available to handicapped individuals for all code test elements at all exam sessions [97.509(k)].

The 13 and 20-WPM code test credit will be granted only to individuals who have passed the 5-WPM code test [Element 1(A)]. There are no exemptions for the basic 5-WPM code test other than the exception for commercial radiotelegraph license holders, described above [97.505(a)(7)].

License Exams

Teams of three accredited amateur **Volunteer Examiners (VE)**, under the supervision of umbrella organizations known as **Volunteer Examiner Coordinators (VEC)**, administer exams for all license applicants at locations across the country at convenient times specified by the VEs, following advance public notice [97.509(a)] and [97.519(a)]. Exams are typically given in public halls and schools, and at conventions, hamfests, flea markets and club meetings.

VEs are present and observe candidates throughout the entire exam process. They are responsible for its proper conduct and supervision under instructions from the VEC [97.509(c)]. VEs grade test papers, report results and handle all paperwork with their VEC. The VEC, in turn, serves as the interface between the Volunteer Examiners and the FCC.

Application

Each candidate must present the examiners with a Form 610, with Section I properly completed, on or before the registration deadline if preregistration is required. Otherwise, applicants present the completed Form 610 to the examiners at the session, prior to the beginning of the exam. Candidates are required by the FCC to bring two identification documents, including their original (not a photocopy) license documents and original Certificates of Successful Completion of Examination (CSCE) [FCC Instructions to Examinee, Form 610]. (You may, however, attach a photocopy of your license to the 610.)

Following Instructions

Candidates must follow the instructions given by the examiners [97.511]. The examiners will immediately terminate the exam if the candidate doesn't follow directions [97.509(c)]. VEs must accommodate an examinee whose physical disabilities require a special exam procedure [97.509(k)].

Code Test

The code test elements can be prepared by the examiners or obtained from the VEC, according to the VEC's instructions [97.507(c)]. The person preparing these tests must hold an Amateur Extra class license [97.507(a)] except that Element 1(A) may be prepared by a General or higher class VE. The test must prove the applicant's ability to send and receive the code at the prescribed speeds. Tests must include all of the following: All letters of the alphabet, numerals 0-9, period, comma, question mark, slant mark and prosigns AR, BT and SK [97.503(a)]. Each letter, numeral, punctuation mark and prosign must be contained at least once in a test message [97.507(d)].

The sending test is no longer required, as the FCC has found that those who can receive the code can send it as well. At their discretion, however, VEs may also administer a sending test [97.509(g)].

The style of test—fill-in-the-blank, multiple-choice, straight-copy—is chosen by the VEC or in some cases, the VEs themselves. In cases where "straight copy" testing is used, each five letters of the alphabet count as one word; each numeral, punctuation mark and prosign counts as two letters of the alphabet [97.507(d)].

Written Test

The formula for creating the written tests is determined by the FCC [97.503(c)].

The VEC selects questions for each test from the appropriate pool of questions approved by the majority of VECs (see "Question Pools" below) [97.507(b)], [97.503(c)] and [97.523]. The VEC keeps the exact exam designs secret, but the general question pools are available to the public (copies of each element's pool are available from ARRL HQ) [97.523].

VEs receive test papers from their VEC, or in some cases create their own and administer them to the candidates in accordance with the VEC's instructions [97.507(c)].

When You Pass

When an applicant passes an exam element, the examiners issue a *Certificate of Successful Completion of Examination* (CSCE) [97.509(l)]. This certificate is required for already-licensed applicants operating with newly acquired privileges of a class higher than that of their permanent license. With a CSCE showing a successful license upgrade, the applicant can (with appropriate identification—see "Interim Identification," later in this chapter) operate with the new privileges prior to the FCC granting the upgraded ticket. The certificate also carries with it a 365-day credit for elements passed when taking subsequent exams.

When the candidate passes all of the exam elements for a license, the examiners indicate that fact on Form 610. Within 10 days following the exam, the examiners must send the application of a successful candidate to the VEC [97.509(m)]. The VEC screens the data before sending it on, within 10 days of receipt, to the FCC Licensing Division in Gettysburg [97.519(b)]. Each VEC must make any examination records available to the FCC upon request. Since January 1995, VECs have been able to file examination data with the FCC electronically. This has sped up the licensing process dramatically. VECs who file electronically must keep the Form 610 documents available for FCC inspection for 15 months [97.519(b),(c)].

Retesting Waiting Period

There is no waiting period before reexamination after failing an exam element. The FCC requires, however, that VECs not use the same set of exam questions in successive exam sessions to ensure that "retest" doesn't mean "remembering" [97.509(f)].

Most VEs will retest candidates immediately if more than one version of the failed elements(s) is available at the session.

Special FCC Exams

At any time, the FCC may readminister any exam given by a VEC, itself or under the supervision of VEs it designates. The FCC can cancel the license of a licensee who doesn't show up for, or fails to pass, an FCC directed reexamination. This allows the FCC to spot check and maintain integrity in the exam process [97.519(d)].

Volunteer Examiners

To be accredited by a VEC as a Volunteer Examiner, you must be 18 years old [97.509(b)(2)]. If your license has ever been suspended or revoked, the FCC will not accept your services as a VE [97.509(b)(4)]. VEs may not give exams to spouses, children, grandchildren, stepchildren, parents, grandparents, stepparents, brothers, sisters, stepbrothers, stepsisters, aunts, uncles, nieces, nephews or in-laws [97.509(d)].

Tests for Novice, Technician and Technician-Plus licenses can be given by VEs holding General, Advanced or Amateur Extra class licenses. Exams for General or higher class elements must be given by VEs who hold Amateur Extra class licenses only [97.509(b)(3)].

Volunteer Examiner Coordinators

VECs are umbrella organizations that serve as interfaces between the Commission and the Volunteer Examiners in the field. They coordinate the efforts of VEs, print and distribute exam papers to VEs and forward successful applicants' Form 610s to the FCC's licensing division [97.519].

Volunteer Examiner Coordinators have entered into special agreements with the FCC after having met certain qualifications: A VEC must be an organization that exists for the purpose of furthering the interests of Amateur Radio [97.521(a)]. It must be able to serve at least one of the VEC regions [listed in Appendix 2 of the Rules] and agree to coordinate test sessions for all classes of license [97.521(b),(c)].

It's the VEC's legal responsibility to accredit a broad range of hams to be Volunteer Examiners, regardless of race, sex, religion or national origin. A VEC may not refuse to accredit a volunteer on the basis of membership (or lack thereof) in an Amateur Radio organization, nor on the basis of the person accepting or declining to accept reimbursement [97.525(b)]. A VEC must not accredit a volunteer to be an examiner, however, if (1) he or she does not meet the VE qualifications; (2) the FCC refuses to accept the services of the volunteer; (3) the VEC refuses to accept the services of the volunteer; (4) the VEC determines that the volunteer is not competent to perform the function of a VE; or (5) the VEC determines that questions of the volunteer's integrity or honesty could compromise the exam [97.525(a)].

Conflict of Interest

The FCC Rules formerly prohibited amateurs from holding a significant interest in, or being an employee of, a company that made or distributed amateur equipment, or published or distributed amateur licensing study materials, unless that amateur could show he or she wasn't engaged in making, publishing or distributing such equipment or materials. Similar conflict of interest requirements applied to VECs. The Telecommunications Act of 1996 dropped all conflict of interest requirements, and the FCC has modified its rules accordingly. The ARRL VEC's famous "Chinese Wall" may go the way of the Berlin Wall.

Interim Identification

If you already hold a license, and qualify for a new higher-class license, you may use your new privileges immediately [97.9(b)]. The FCC has established temporary identifier codes for each class of license. This code is a slant bar (/) followed by a two-letter group: KT for Technician or Technician-Plus classes, AG for General class, AA for Advanced class, and AE for Amateur Extra class. The ID code is usually shown on the certificate issued to you when you successfully complete your exam. It must be added as a suffix to your call sign when you operate with your new privileges [97.119(e)]. For example, a Novice who passes the Technician written test [Element 3(A)], may operate on 2 meters immediately after receiving the CSCE, but would have to give his or her call sign as "WX4XXX/Temporary KT" until the FCC grants the upgrade. If the amateur operates on the Novice frequencies, he or she does not have to add the special suffix.

Question Pools

Question pools are large numbers of exam questions arranged by topic from which specific question sets for tests are extracted [97.3(a)(32),(33)]. Pools for each written test element are developed by all VECs working together at periodic question-pool review sessions [97.523].

Amateur Extra class VEs are allowed to prepare questions for the pool, and question sets for specific exams, for any element. Advanced class VEs may prepare questions for Novice [Element 2], Technician and Technician-Plus [Element 3(A)], and General class [Element 3(B)] written exams only. General class VEs may prepare questions for the written Novice [Element 2], Technician and Technician-Plus class tests [Element 3(A)] only [97.507(a)].

To keep question pools current, they are changed in accordance with a published schedule. The pools must contain at least ten times the number of questions required for a single exam. For example, the Advanced-class test consists of 50 questions, so the pool must have at least 500 questions available. Thus, although a candidate can see the exact questions that *might* appear on his exam, he won't know which 50 will show up [97.523].

The question pools must be published and made available to the public prior to their use in making a question set. No question may appear on an exam unless it appears in the appropriate, current question pool [97.523].

Examination Expense Reimbursement

The FCC allows VEs and VECs to charge exam fees to recover their out-of-pocket expenses incurred in preparing, processing and administering exams [97.527(a)]. There is a ceiling on fees, however, which is announced annually by the FCC. The basis for the maximum fee allowable is $4 for 1984, adjusted annually each January 1 thereafter for changes in the Consumer Price Index [97.527(b)].

The Telecommunications Act of 1996 eliminated the requirement for VEs who collect examination fees to maintain records of expenses and reimbursements. The FCC Rules no longer require VEs to file written certification of expenses incurred and reimbursements collected with their VEC, and in turn the VECs no longer have to file them with the FCC.

VE Conduct

No VE may administer or certify any exam by fraudulent means or for monetary or other gift, including reimbursement in excess of the permitted amount.

Postscript: 80 Years Of Licensing

Until 1912, there was no licensing, no regulations and no governing body to oversee "wireless" activities on the airwaves. Before long, however, it became evident to federal authorities that regulation was needed to maintain order: Conflicts between amateur stations and those used by the Navy and commercial services were on the increase. The first regulation came in the form of licensing.

The era of mandatory licensing began when the US Department of Commerce and Labor, under the authority of the Radio Act of 1912, created the Amateur First Grade and Amateur Second Grade operator licenses. The two classes bestowed identical privileges and, at least theoretically, required identical qualifications.

Amateur First Grade applicants took written tests on radio laws, regulations, and the proper adjustment and operation of equipment. The code sending and receiving tests, originally 5 WPM, increased to 10 WPM by 1919. Candidates for Amateur Second Grade, in contrast, certified to Radio Inspectors by mail that they could meet these requirements, but were unable to attend an examination.

Until 1933, station and operator licenses were issued as separate, diploma-sized-certificates. The type of station license held (originally General, Special or Restricted) determined permissible operating wavelengths and power.

In 1923, the Department of Commerce created the Amateur Extra First Grade, a license so special it was printed on pink paper! Only Amateur Extra First Grade licensees thereafter qualified for "Special" station licenses, which had distinctive call signs and conveyed CW privileges on wavelengths longer than 200 meters.

Qualifications for the new class included two years' experience as a licensed operator and a written examination that, among other items, required the applicant to diagram a transmitter and receiver and then explain their principles of operation. The code tests were given at 20 WPM, the speed required of Commercial First Class operators.

As amateur interests shifted to shortwaves, the Amateur Extra First Grade's popularity declined. Only six such licenses were issued in 1926 and the class was discontinued the following year. Reinstated in 1928 with new privileges (described below) added in 1929, the class attracted several hundred licensees most years until its permanent deletion in 1933.

The Radio Act of 1927 transferred the power to issue station licenses to the Federal Radio Commission (FRC) while preserving the authority of the Commerce Department's Radio Division to issue operator licenses. Months later, the Radio Division redesignated the Amateur First and Second Grade classes as Amateur Class and Temporary Amateur, respectively. To First Grade licensees, the change meant little more than a new name. Temporary Amateur differed from the previous Second Grade, however, in that the former expired in one year and (after 1932) could not be renewed. Hams could no longer indefinitely avoid taking an examination.

In late 1929, the Radio Division began endorsing Amateur Extra First Grade licenses for "unlimited radiotelephone privileges." Initially, the endorsement authorized voice privileges on the 20-meter band. In 1932, the endorsement became available to other amateurs having at least one year of experience, upon passing a special test on radiotelephone subjects. At the same time, phone use of 75 meters was also reserved to holders of endorsed licenses.

The Radio Division merged with the FRC in 1932. A year later, the FRC completely revised the amateur regulations. Station and operator licenses were thereafter combined on a single, wallet-sized card.

The Amateur's basic license was endorsed as Class A, B or C. All three classes required code tests at 10 WPM (13 WPM after 1936). Class A conveyed exclusive phone use on 20 and 75 meters. It required one year of prior experience and a written examination on radiotelephone and radiotelegraph theory and amateur regulations.

Classes B and C conveyed all privileges other than those reserved to Class A. The written test for those classes was less comprehensive than that for Class A with regard to radiotelephone theory. The two classes differed in that Class C written examinations were furnished by mail to applicants residing at least 125 miles from the nearest FRC quarterly examining point. Class C code tests were administered by Class A and B licensees acting as volunteer examiners.

Amateur Extra First Grade licensees qualified for Class A privileges upon renewal. Amateur Class licensees were grandfathered into Class B. Temporary Amateur licenses could not be renewed, however, so holders of this class had to qualify anew in Class B or C upon expiration of their licenses.

The Federal Communications Commission (FCC) succeeded the FRC with the passage of the Communications Act of 1934. It revised the regulations in 1951 to create the license-class names that are familiar today. Advanced, General and Conditional licenses replaced Classes A, B and C, respectively. The Advanced class was closed to new applicants

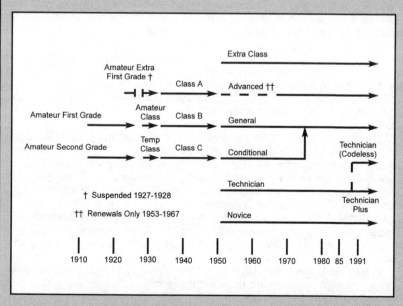

The evolution of Amateur Radio licensing over the past eight decades.

in January 1953, although renewal of existing licenses continued. A month later, the 20 and 75-meter "Class A Phone Bands" were opened to General and Conditional licensees.

The same rule-making action created the Amateur Extra, Novice and Technician classes. The Extra Class originally required two years' experience as a Conditional (or Class C) licensee or higher, code tests at 20 WPM and a theory examination more comprehensive than that previously given for Class A. No exclusive privileges were reserved for the new class.

The Technician ticket originally conveyed all amateur privileges above 220 MHz. Novices were initially restricted to CW operation on portions of the 11 and 80-meter bands, and voice at 145-147 MHz, at 75-W input, using only crystal-controlled transmitters. The first Novice licenses expired after one year and could not be renewed. After 1954, Novice and Technician exams were obtained from the FCC by mail and administered through volunteer examiners. In 1976, the system changed again: Potential Technician-class licensees were required to appear before an FCC examiner, although existing Technician licenses were grandfathered.

A glimpse at our current regulations reveals that the licensing system has undergone many changes since 1951. Although it is beyond the scope here to examine them all, some of the most important have been the establishment of license upgrading incentives and the reopening of the Advanced class to new licensees in 1967; elimination of activity and code speed requirements for renewal; the expansion and realignment of Novice and Technician frequency privileges, notably in 1976; the increase of the Novice power level, the removal of the crystal-control requirement and the merger of Conditional licenses into the General class in 1976; the extension of license terms to 10 years in 1983; and Novice Enhancement in 1987.

On February 14, 1991, the FCC removed the code requirement from the Technician-class license, creating the first codeless license class in the US. Ever since the early days, the amateur service has been in a constant state of evolution, and there is every reason to believe that Amateur Radio of the future will look quite different.—*Neil D. Friedman, N3DF*

Violations may result in revocation or suspension of the VE's license [97.509(e)].

The All-Purpose Form 610

The FCC Form 610 is the Commission's multipurpose amateur licensing form. It's the only official contact most amateurs have with the FCC. Amateurs use it to apply for a new license, an upgraded license, license renewal, modification or reinstatement [97.17(b)] and [97.21(a)]. See Fig 1.

Five versions of the 610 exist for different purposes. For most, the standard FCC Form 610 is used. Foreign amateurs applying for permission to operate in the US use FCC Form 610-A [97.17(b)]. Club and military recreation station trustees use Form 610-B to apply for new station licenses, and for renewal and modification purposes. The Commission will not issue new RACES station licenses, but trustees may renew or modify existing RACES licenses using Form 610-B [97.19(b)]. As of December 20,

FEDERAL COMMUNICATIONS COMMISSION
GETTYSBURG, PENNSYLVANIA

Approved by OMB
3060-0003
Expires 8/31/96
See instructions for
information regarding
public burden estimate.

APPLICATION FORM 610 FOR
AMATEUR OPERATOR/PRIMARY STATION LICENSE

SECTION 1 - TO BE COMPLETED BY APPLICANT (See instructions)

1. Print or type last name | Suffix | First name | Middle initial | 2. Date of birth
   month / day / year

3. Mailing address (Number and street) | City | State code | ZIP code

4. I HEREBY APPLY FOR (make an X in the appropriate box(es)):

4A. ☐ **EXAMINATION** for a new license

4B. ☐ **EXAMINATION** for upgrade of my operator license class

4C. ☐ **CHANGE** my name on my license to my new name in Item 1. My former name was:

___ (Last name) ___ (Suffix) ___ (First name) ___ (MI)

4D. ☐ **CHANGE** my mailing address on my license to my new address in Item 3

4E. ☐ **CHANGE** my station call sign systematically (See instructions) Applicant's Initials _____

4F. ☐ **RENEWAL** of my license

5. Unless you are requesting a new license, attach the original or a photocopy of your license to the back of this Form 610 and complete Items 5A and 5B. | 5A. Call sign shown on license | 5B. Operator class shown on license

6. Would an FCC grant of your request be an action that may have a significant environmental effect? ☐ NO ☐ YES (Attach required statement)

7. If you have filed another Form 610 that we have not acted upon, complete Items 7A and 7B. | 7A. Purpose of other form | 7B. Date filed
month / day / year

WILLFUL FALSE STATEMENTS MADE ON THIS FORM ARE PUNISHABLE BY FINE AND/OR IMPRISONMENT, (U.S. CODE, TITLE 18, SECTION 1001), AND/OR REVOCATION OF ANY STATION LICENSE OR CONSTRUCTION PERMIT (U.S. CODE, TITLE 47, SECTION 312(A)(1))) AND/OR FORFEITURE (U.S. CODE, TITLE 47, SECTION 503).

I CERTIFY THAT ALL STATEMENTS AND ATTACHMENTS ARE TRUE, COMPLETE, AND CORRECT TO THE BEST OF MY KNOWLEDGE AND BELIEF AND ARE MADE IN GOOD FAITH; THAT I AM NOT A REPRESENTATIVE OF A FOREIGN GOVERNMENT; THAT I WAIVE ANY CLAIM TO THE USE OF ANY PARTICULAR FREQUENCY REGARDLESS OF PRIOR USE BY LICENSE OR OTHERWISE; AND THAT THE STATION TO BE LICENSED WILL BE INACCESSIBLE TO UNAUTHORIZED PERSONS.

8. Signature of applicant (Do not print, type, or stamp. Must match name in Item 1.)

✖ _____

( ) _____
Daytime Telephone Number

9. Date signed
month / day / year

SECTION 2 - TO BE COMPLETED BY ALL ADMINISTERING VE's

A. Applicant is qualified for operator license class:

☐ NOVICE (Elements 1(A), 1(B), or 1(C) and 2)
☐ TECHNICIAN (Elements 2 and 3(A))
☐ TECHNICIAN PLUS (Elements 1(A), 1(B), or 1(C), 2 and 3(A))
☐ GENERAL (Elements 1(B) or 1(C), 2, 3(A) and 3(B))
☐ ADVANCED (Elements 1(B) or 1(C), 2, 3(A), 3(B) and 4(A))
☐ AMATEUR EXTRA (Elements 1(C), 2, 3(A), 3(B), 4(A) and 4(B))

B. VEC receipt date:

C. Name of Volunteer-Examiner Coordinator (VEC):

D. Date of VEC coordinated examination session: | E. Examination session location:

I CERTIFY THAT I HAVE COMPLIED WITH THE ADMINISTERING VE REQUIREMENTS IN PART 97 OF THE COMMISSION'S RULES AND WITH THE INSTRUCTIONS PROVIDED BY THE COORDINATING VEC AND THE FCC

1st VE's name (Print First, MI, Last, Suffix)	VE's station call sign	VE's signature (must match name)	Date signed
2nd VE's name (Print First, MI, Last, Suffix)	VE's station call sign	VE's signature (must match name)	Date signed
3rd VE's name (Print First, MI, Last, Suffix)	VE's station call sign	VE's signature (must match name)	Date signed

FCC Form 610
March 1995

Fig 1—FCC Form 610

1994 the FCC can send out Form 610-R as a renewal short form. The vanity call sign program uses Form 610-V. Because a fee is involved, it is never submitted directly to the FCC's Gettysburg office. Form 610-V is used to apply for a vanity call sign, or to modify or renew a license that shows a vanity call sign.

Use the Latest Edition

Use a current edition of Form 610. As this is written, the FCC will only accept 610s

ATTACH ORIGINAL OR A PHOTOCOPY OF YOUR LICENSE HERE:

SECTION 3 - TO BE COMPLETED BY PHYSICIAN

PHYSICIAN'S CERTIFICATION OF DISABILITY
Please see notice below

Print, type, or stamp physician's name: _____

Street address: _____

City, State, ZIP code: _____

Office telephone number: (___) _____

I CERTIFY THAT I have read the Notice to Physician Certifying to a Disability, and that the person named in Item 1 on the reverse is severely handicapped, the duration of which will extend for more than 365 days beyond this date. Because of this severe handicap, this person is unable to pass a 13 or 20 words per minute telegraphy examination. I am licensed to practice in the United States or its Territories as a doctor of medicine (M.D.) or doctor of osteopathy (D.O.). I have considered the accommodations that could be made for this person's disability and have determined that, even with accommodations, this person would be unable to pass a 13 or 20 words per minute telegraphy examination.

WILLFUL FALSE STATEMENT IS PUNISHABLE BY FINE AND IMPRISONMENT (U.S. CODE TITLE 18, SECTION 1001)

➤ _____

PHYSICIAN'S SIGNATURE (DO NOT PRINT, TYPE, OR STAMP)   M.D. or D.O.   DATE SIGNED

PATIENT'S RELEASE
Authorization is hereby given to the physician named above, who participated in my care, to release to the Federal Communications Commission any medical information deemed necessary to process my application for an amateur operator/primary station license.

➤ _____

APPLICANT'S SIGNATURE (DO NOT PRINT, TYPE, OR STAMP)   DATE SIGNED

NOTICE TO PHYSICIAN CERTIFYING TO A DISABILITY

You are being asked by a person who has already passed a 5 words per minute telegraphy examination to certify that, because of a severe handicap, he/she is unable to pass a 13 or 20 words per minute telegraphy examination. If you sign the certification, the person will be exempt from the examination. Before you sign the certification, please consider the following:

THE REASON FOR THE EXAMINATION - Telegraphy is a method of electrical communication that the Amateur Radio Service community strongly desires to preserve. We support their objective by authorizing additional operating privileges to amateur operators who increase their skill to 13 and 20 words per minute. Normally, to attain these levels of skill, intense practice is required. Annually, thousands of amateur operators prove by passing examinations that they have acquired the skill. These examinations are prepared and administered by amateur operators in the local community who volunteer their time and effort.

THE EXAMINATION PROCEDURE - The volunteer examiners (VEs) send a short message in the Morse code. The examinee must decipher a series of audible dots and dashes into 43 different alphabetic, numeric and punctuation characters used in the message. To pass, the examinee must correctly answer questions about the content of the message. Usually, a fill-in-the-blanks format is used. With your certification, they will give the person credit for passing the examination, even though they do not administer it.

MUST A PERSON WITH A HANDICAP SEEK EXEMPTION?

No handicapped person is required to request exemption from the higher speed telegraphy examinations, nor is anyone denied the opportunity to take the examinations because of a handicap. There is available to all otherwise qualified persons, handicapped or not, the Technician Class operator license that does not require passing a telegraphy examination. Because of international regulations, however, any handicapped applicant requesting exemption from the 13 or 20 words per minute examination must have passed the 5 words per minute examination.

ACCOMMODATING A HANDICAPPED PERSON - Many handicapped persons accept and benefit from the personal challenge of passing the examination in spite of their hardships. For handicapped persons without an exemption who have difficulty in proving that they can decipher messages sent in the Morse code, the VEs make exceptionally accommodative arrangements. They will adjust the tone in frequency and volume to suit the examinee. They will administer the examination at a place convenient and comfortable to the examinee, even at bedside. For a deaf person, they will send the dots and dashes to a vibrating surface or flashing light. They will write the examinee's dictation. Where warranted, they will pause in sending the message after each sentence, each phrase, each word, or each character to allow the examinee additional time to absorb and interpret what was sent. They will even allow the examinee to send the message, rather than receive it.

YOUR DECISION - The VEs rely upon you to make the necessary medical determination for them using your professional judgement. You are being asked to decide if the person's handicap is so severe that he/she cannot pass the examination even when the VEs employ their accommodative procedures. The impairment, moreover, will last more than one year. This procedure is not intended to exempt a person who simply wants to avoid expending the effort necessary to acquire greater skill in telegraphy. "The person requesting that you sign the certification will give you names and addresses of VEs and other amateur operators in your community who can provide you with more information on this matter.

DETAILED INSTRUCTIONS - If you decide to execute the certification, you should complete and sign the Physician's Certification of Disability on the person's FCC Form 610. You must be an M.D. or D.O. licensed to practice in the United States or its Territories. The person must sign a release permitting disclosure to the FCC of the medical information pertaining to the disability.

FCC Form 610
March 1995

dated November 1993 or later. Check the date on your form or it may be returned!

The Physician's Certification of Disability form appears on the back of the 610. It is used by persons with severe disabilities seeking waivers of the 13 and 20-WPM code requirement.

To obtain a new Form 610, 610-A, 610-B or 610-V, send a business-sized, self-addressed, stamped envelope to ARRL, 225 Main St, Newington, CT 06111. The 610-R is sent out only by the FCC.

License Renewals

The FCC requires that you apply for a renewal no more than 90 days in advance of your expiration date. If your renewal arrives at the FCC's Gettysburg office prior to the expiration date of your license, you may continue to operate while waiting for the renewal. The FCC is authorized to send the Form 610-R to an amateur whose license is about to expire (not available at press time). You may still use the normal 610 to renew, and should do so if you don't receive the 610-R at least 30 days before your expiration date. You must attach the original or a photocopy of your license if you use the 610. You do not need to do so if you use the 610-R [97.21].

If you let your license lapse, you have a two-year grace period to request that it be reinstated. During this grace period, you are not permitted to operate until your application for reinstatement is granted. It will be backdated to the date of your license's expiration

The end of the line for your 610 is the FCC Licensing Division in Gettysburg, Pennsylvania.

Your Form 610 arrives at the FCC either in paper form or electronically from a VEC. It is immediately forwarded to the amateur processors. Here, every Form 610 that is submitted to the FCC is checked for validity by a professional staff equivalent to only 2¹/₂ people!

Once your Form 610 passes muster, your record is updated on the FCC computer. Your new license is effective as soon as it is entered into the FCC's database.

Problems or questions concerning a license? The Customer Assistance Branch is the public's contact point with the FCC Licensing Division. People seated at terminals along the wall are not FCC staff but visitors who are researching questions.

[97.21(b)]. Licenses are no longer automatically renewed when a modification is made.

The normal term for a license is 10 years [97.25(a)]. Reciprocal permits are not renewable and are issued for a one-year term [97.25(b)]. Permittees may, however, apply for a new permit each year.

Your Mailing Address

You must always put your mailing address on an application. This address must be in an area regulated by the FCC. For most of us, that means in the US; see Appendix 1 to Part 97. The mailing address must be one where you can receive mail via the US Postal Service—a street address or post office box, for example. The FCC no longer keeps a record of your station location.

If your mailing address changes, the FCC requires that you file a timely application for license modification. If correspondence sent by the FCC to your mailing address is returned because you failed to provide a correct address, your license can be suspended or revoked. This applies to both licenses and reciprocal permits [97.23].

The Gettysburg Address

All applications except registrations for VEC-sponsored exams or vanity call sign applications should be sent to FCC, 1270 Fairfield Rd, Gettysburg, PA 17325-7245 [97.17(c) and (d); 97.21(a)]. Except for vanity call sign applications, there are no fees for filing applications for license renewal or modification with the FCC. You may have to pay to take a VEC-sponsored examination.

If you have not received a response from the FCC concerning your application within 90 days, write to the FCC in Gettysburg. Include with your letter a photocopy of your application or the following information: Name and address, birth date, present call sign and class of license (if any), date of application, and VEC. You may also call the FCC's Customer Assistance Branch at 800-322-1117 weekdays from 8 AM to 4:30 PM Eastern time.

Lost or Destroyed Licenses

To get a replacement for a lost, mutilated or destroyed license, mail a letter explaining the situation to the FCC in Gettysburg. They'll provide you with a duplicate license with the same expiration date as the lost/destroyed license [97.29].

Call Signs

If ever there was a subject whose significance was more apparent to radio amateurs than to nonhams, it is call signs. There's an emotional dimension to that set of letters and numerals by which we're identified that is not fully understood by unlicensed people.

Back in 1978, the FCC closed a call sign assignment program where eligible Amateur Extra class licensees could request and be granted special consideration in call sign selection. The FCC went to the present system of computer-assigned call signs (see "Q&A Call Signs"), where you become eligible (at your option) for progressively more "desirable" (i.e., shorter) call signs as you progress up the licensing ladder.

The Vanity Call Sign System

In 1993 the US Congress authorized the FCC to establish a vanity call sign

Federal Communications Commission
1270 Fairfield Road
Gettysburg, PA 17325-7245
(800) 322-1117 or (717) 337-1212

APPLICATION FOR AN AMATEUR CLUB, RACES OR MILITARY RECREATION STATION LICENSE

BE SURE TO READ INSTRUCTIONS ON REVERSE SIDE. PLEASE PRINT OR TYPE.

➡ ATTACH PRESENT CLUB, RACES OR MILITARY RECREATION LICENSE OR PHOTOCOPY THEREOF TO THE BACK OF APPLICATION. IF THE LICENSE HAS BEEN LOST OR DESTROYED, ATTACH EXPLANATION.

1. CLUB, RACES OR MILITARY RECREATION STATION CALL SIGN:	2. NAME OF CLUB STATION TRUSTEE OR LICENSE CUSTODIAN: (Last, Suffix, First, MI)	3. DATE OF BIRTH:
		MONTH DAY YEAR

4. CURRENT MAILING ADDRESS: (Number and Street, City, State, ZIP Code)

5. APPLICATION IS FOR (Check one):

☐ NEW   ☐ MODIFICATION   ☐ RENEWAL

6. TRUSTEE'S PRIMARY STATION CALL SIGN:	7. NAME OF CLUB, RACES ORGANIZATION OR MILITARY RECREATION ENTITY:

8. WOULD AN FCC GRANT OF THIS REQUEST BE AN ACTION THAT MAY HAVE A SIGNIFICANT ENVIRONMENTAL EFFECT?

☐ NO   ☐ YES (Attach required statement)

9. APPLICANT CLASSIFICATION:

☐ CLUB   ☐ MILITARY RECREATION   ☐ RACES

CERTIFICATION

RESPONSIBLE OFFICIAL: I certify that the above named person is the station trustee or license custodian authorized to apply for and hold an amateur radio station license for this organization, society or entity.

WILLFUL FALSE STATEMENTS MADE ON THIS FORM ARE PUNISHABLE BY FINE AND/OR IMPRISONMENT (US CODE, TITLE 18, SECTION 1001), AND/OR REVOCATION OF ANY STATION LICENSE OR CONSTRUCTION PERMIT (US CODE, TITLE 47, SECTION 312(A)(1)), AND/OR FORFEITURE (US CODE, TITLE 47, SECTION 503).

10. SIGNATURE: (Must not agree with Item 2)	(Title or authority to approve)	DATE SIGNED:

APPLICANT: I certify that all statements herein and attachments herewith are true, complete and correct to the best of my knowledge and belief and are made in good faith; that I am not the representative of a foreign government; that I waive the claim to the use of any particular frequency regardless of prior use by license or otherwise; that the station to be licensed will be inaccessible to unauthorized persons.

WILLFUL FALSE STATEMENTS MADE ON THIS FORM ARE PUNISHABLE BY FINE AND/OR IMPRISONMENT (US CODE, TITLE 18, SECTION 1001), AND/OR REVOCATION OF ANY STATION LICENSE OR CONSTRUCTION PERMIT (US CODE, TITLE 47, SECTION 312(A)(1)), AND/OR FORFEITURE (US CODE, TITLE 47, SECTION 503).

11. APPLICANT'S SIGNATURE: (Must agree with Item 2)	DATE SIGNED:	DAYTIME TELEPHONE NUMBER:

FCC 610B
March 1995

Fig 2—FCC Form 610-B

system. In early 1995, the FCC adopted new rules implementing such a system by which any amateur, not just Amateur Extra licensees, may request a call sign of choice. Applicants can list up to 25 possible call signs on Form 610-V. See Fig 4. The FCC is authorized by Congress to charge a fee to offset their costs for administering the program. The completed 610-V must be sent, with the $30 fee ($3/yr for 10 years), to: Federal Communications Commision, Amateur Vanity Call Sign Requests, POB 358924, Pittsburgh, PA 15251-5924. Do not send Form 610-V to the FCC's Gettysburg office [97.19(b)]. If none of the requested call signs are available, applicants will retain their original call sign. Call signs will be issued on a first-come, first-served basis.

Amateurs can request previously held call signs, call signs held by deceased close relatives, or any other vacant call sign from the sequential call sign system. Applicants are not limited to call signs from their call district, but may request one from any call district. The applicant may request call signs only from the appropriate call sign group for his or her license class, or a lower one. Applicants may request an available call sign from any call area, except that only applicants with a mailing address in the specific territory may request a call sign reserved for Alaska, Hawaii, American Samoa, Marianas, Guam, Puerto Rico or the US Virgin Islands. This limitation does not apply to applicants requesting a previously held call sign, that of a close relative, or of a deceased club member (with a relative's permission) [97.19(d)(4)].

Call signs vacated for whatever reason are not available to the vanity call sign system for two years. This does not apply to a close relative or to a club applying with the written permission of a close relative. A close relative is defined as a spouse, child, grandchild, stepchild, parent, grandparent, stepparent, brother, sister, stepbrother, stepsister, aunt, uncle, niece, nephew, or in-law [97.19(c)(3)].

A club wishing to apply for a vanity call sign must first have a regular club license. Clubs without a station license must first apply for one by sending a Form 610-B to the FCC's Gettysburg office. See Fig 2. There is no fee for a basic club station license. The license they receive will have a sequentially issued call sign from Group D (Novice). Once the club station license is in hand, they may apply for a vanity call sign using Form 610-V accompanied by the required fee. Clubs are limited to vanity call signs from the appropriate call sign group for the station trustee's license class, or a lower one. Military recreation and RACES stations are not eligible to apply for vanity call signs [97.14(a)].

FCC Modifications to Your License

The FCC reserves the right to modify your license at any time if it determines that such action is in the public interest, convenience and necessity. The FCC, however, must give you the opportunity to show why your license should not be modified. If you don't respond adequately or don't take an active interest in the issue, your license and operation may be restricted temporarily or permanently [97.27].

Club Stations

In 1978, as an economy measure, the FCC proposed to eliminate the licensing of club stations. It didn't simply want to stop issuing new club licenses, but renewals as well! This set off a storm of protest from clubs, many of which had held the same call since before the FCC itself was created. The idea of not renewing existing licenses was dropped, but no new club licenses were issued between 1978 and 1995. Military recreation stations, RACES stations and short-term special-event stations fell victim to the same budget-cutting ax. For a long time club, military recreation and RACES station records weren't even in the Commission's computer data base, but were maintained manually on file cards.

On March 24, 1995, the FCC began once again accepting applications for new club and military recreation station licenses [97.17(h)]. New station licenses are issued call signs from call sign Group D (Novice). Club stations are eligible to

Q&A—Call Signs

Q. How does the FCC's sequential call sign system work?

A. Refer to Figs A and B. Call signs are issued sequentially by the FCC computer in Gettysburg [97.17(f)]. The computer selects call signs from "blocks" or "groups" of call signs that reflect the licensee's operator class, mailing address and call sign region. The FCC issues call signs from four groups. Refer to Fig B. **Group A** call signs are issued only to Amateur Extra class operators. **Group B** call signs are assigned to amateurs holding Advanced class licenses. **Group C** call signs are issued to Technician, Technician Plus and General class operators. **Group D** call signs go to Novice class operators and new club and military station licenses.

The FCC sequential call sign assignment system selects only call signs that have not been previously issued. This is the system that automatically issues calls to new licensees and to persons who check and initial item 4E on Form 610. The system assigns call signs from the next lower group after all call signs from one group have been depleted. This has already happened in several areas. For example, there are no more 1×2 or 2×1 call signs issued to Amateur Extra class hams by the sequential system. In addition, all Group C call signs have been assigned in many districts, so applicants in those districts receive Novice call signs.

Starting in 1995, the FCC once again began honoring requests for specific call signs. The vanity call sign system allows any amateur to request a specific call sign. The requested call sign must not be currently

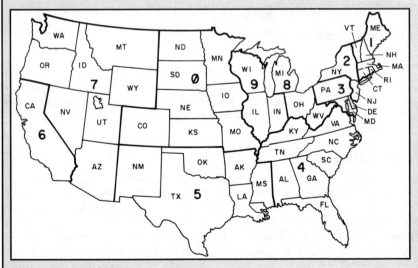

Fig A—Continental US call districts.

assigned, and must be available from the sequential call sign system [97.19].

Q. I would like to know what the most recent call sign issued in my district is. Can this information be obtained?

A. Yes. At the beginning of each month, the FCC releases a public notice that lists the latest call signs issued in each group. The FCC's Gettysburg Consumer Assistance Branch maintains a "call sign hotline" for obtaining this information, 800-322-1117, or 717-337-1212. The same information is available from the ARRL's Regulatory Information Branch. This information is regularly carried in W1AW bulletins and printed in the *ARRL Letter* and in the Happenings column of *QST*.

Q. When am I eligible for a new call sign?

A. You are eligible for a new call sign at any time. Most amateurs, however, request a new call sign only if they upgrade their operator license class or change their mailing address to a different call sign area.

Q. When I upgrade, can I keep my present call sign?

A. Yes, you can keep your present call sign if you want to. Call signs are changed only when you specifically request it by checking and initialing item 4E on the FCC Form 610.

Q. A local ham recently became a Silent Key. Will his call sign automatically be deleted from FCC records?

A. The FCC no longer deletes the call signs of Silent Keys unless cancellation is specifically requested. Because the term of an amateur license is now 10 years, the family of the deceased amateur could continue to receive unwanted correspondence for quite some time. To avoid this, and make the call sign available to the vanity call sign system, the family of the deceased amateur should return the license to the FCC or, in the event that the license can't be found, a letter should be sent to the FCC stating that the amateur is deceased and requesting cancellation of the license.

Q. Fifteen years ago I held the call sign W4AAA, but I let my license lapse. I recently passed my Technician examination and I received the call sign N4XXX. Is there any chance that I may regain my old call sign of W4AAA?

A. Yes, by way of the FCC's vanity call sign system. Using FCC Form 610-V, you may apply for a previously held call sign as long as it is not currently held by another operator.

Amateur Station Sequential Call Sign System

A single unique call sign is assigned to each amateur station. A station is reassigned the same call sign upon renewal or modification of its license, unless the licensee applies for a change to a new call sign on FCC Form 610. Each new call sign is sequentially selected from the alphabetized regional-group list for the licensee's operator class and mailing address.

Each call sign has a one-letter prefix (K, N, W) or a two-letter prefix (AA-AL, KA-KZ, NA-NZ, WA-WZ) and a one, two, or three-letter suffix separated by a numeral (1-∅) indicating the geographic region. (The numeral ∅ indicates the tenth call area). Certain combinations of letters are not used. When the call signs in any regional-group list are exhausted, the selection is made from the next lower group. The groups are:

Group A. For Primary stations licensed to Amateur Extra class operators.

Call areas 1 through ∅: Prefix K, N or W, and two-letter suffix; two letter prefix with first letter A, K, N or W, and one-letter suffix; two-letter prefix with first letter A, and two-letter suffix.
Region 11: Prefix AL, KL, NL, or WL, and one-letter suffix.
Region 12: Prefix KP, NP, or WP, and one-letter suffix.
Region 13: Prefix AH, KH, NH, or WH, and one-letter suffix.

Group B. For Primary stations licensed to Advanced class operators.

Call areas 1 through ∅: Two-letter prefix with first letter K, N or W, and two-letter suffix.
Region 11: Prefix AL, and two-letter suffix.
Region 12: Prefix KP, and two-letter suffix.
Region 13: Prefix AH, and two-letter suffix.

Group C. For Primary stations licensed to General, Technician, and Technician Plus class operators.

Call areas 1 through ∅: Prefix K, N, or W, and three-letter suffix.
Region 11: Prefix KL, NL, or WL, and two-letter suffix.
Region 12: Prefix NP or WP, and two-letter suffix.
Region 13: Prefix KH, NH, or WH, and two-letter suffix.

Group D. For Primary stations licensed to Novice class operators, and for club and military recreation stations.

Call areas 1 through ∅: Two-letter prefix with first letter K or W, and three-letter suffix.
Region 11: Prefix KL or WL, and three-letter suffix.
Region 12: Prefix KP or WP, and three-letter suffix.
Region 13: Prefix KH or WH, and three-letter suffix.

Region 11. Alaska. The numeral is 1 through ∅. (KL9KAA-KL9KHZ is reserved for assignment to U.S. personnel stationed in Korea.)

Region 12. Caribbean Insular areas. The numeral 1 indicates Navassa Island; 2 indicates Virgin Islands; 3 or 4 indicates Commonwealth of Puerto Rico except Desecheo Island; and 5 indicates Desecheo Island.

Region 13. Hawaii and Pacific Insular areas. The numeral 1 indicates Baker or Howland Island; 2 indicates Guam; 3 indicates Johnston Island; 4 indicates Midway Island; 5 indicates Palmyra or Jarvis Island; 5 followed by suffix letter K indicates Kingman Reef; 6 or 7 indicates Hawaii except Kure Island; 7 followed by the letter K indicates Kure Island; 8 indicates American Samoa; 9 indicates Wake, Wilkes or Peale Island; and ∅ indicates the Commonwealth of Northern Mariana Islands.

From FCC Fact Sheet #206 (PR-5000), February 1995.

Fig B—Call Sign Groups.

History of Call Signs At A Glance

1995	Vanity Call Sign Program
1980	
AA2Z KJ4KB N1EER KA1SIP	Present call sign assignment system adopted; call signs reflect license class.
N1ER AA4AT	FCC opens new blocks for Extra Class; drops 25-year requirement. First N 1 × 2 and A 2 × 2 call signs.
WD4AAA WC1AAA	New hams receive WD prefixes in some areas. RACES calls issued with WC prefixes.
WR1AUJ	WR calls issued to repeaters.
1970	
W4AA	FCC allows Extras licensed more than 25 years to request one-by-two calls.
WB6AAA	First WB prefixes appear.
1960	
WA2AAE WV2AYO	WN calls run low in second and sixth call areas; FCC issues WV prefixes to Novices. First WA prefixes appear.
WN4TYU	Novice class license and special call signs adopted.
1950	
K9AAA	Atlantic City conference of 1947 assigns AA to AL block to US, but the block was not turned over for amateur use until later. K-prefix calls appear in continental US.
KV4AA	Call areas redrawn to current status. US territories, posessions receive special two-letter prefixes.
1940	
1930	
W1AW	Washington Conference of 1927 assigns international prefixes.
nu6ol	"International intermediates" used to reduce confusion in international QSOs.
1920	
1ANA	First three-letter suffixes appear in 1914 *Call Book*.
1WH	With the passage of the Radio Act of 1912, hams received their first call signs: a number followed by two letters.
SNY	Prior to 1912, there was no licensing. Pioneers made up their own calls. Hiram Percy Maxim used SNY in 1911.
1910	

Q&A—Club Stations

Q. Our new club, the Rochester Radio Ruminators, wants to set up a station at our club headquarters. How do we obtain a club station license?

A. Designate a responsible member of your group as station trustee. If possible this person should be an Amateur Extra class licensee so that all classes of license may use the station without special identification procedures. You must send a completed FCC Form 610-B to the Gettysburg office. As always, the station trustee and the control operator are responsible for the proper operation of the station [97.103(a)].

Q. Our club holds a club station license. We need to change our station trustee and our mailing address. How do we go about modifying our license?

A. You'll need a Form 610-B entitled "Application For Amateur Club Or Military Recreation Station License." This is one of several 610 forms. To change your mailing address for the club, or to make any other modification, the station trustee must send the completed Form 610-B to FCC, 1270 Fairfield Rd, Gettysburg, PA 17325-7245. When a new trustee in named, he or she fills out and submits the form. In all cases, a club officer other than the trustee must also sign the form certifying that the applicant has been appointed station trustee. The FCC will modify the license and send it back to the trustee in about a month [97.19(a)].

apply for a vanity call sign through the vanity call sign program; RACES and military recreation stations are not [97.19(a)].

Foreign Hams Visiting the US

The US receives thousands of foreign amateur visitors every year, and it is only natural that they want to experience the thrill of operating here. The US has entered into "reciprocal operating agreements" with other nations to allow visiting operator privileges. Only amateurs from countries with which the US has signed a reciprocal operating agreement are eligible to obtain a reciprocal operating permit. Reciprocal agreements are negotiated by the US Department of State and the foreign ministry of the other country, not the FCC. In general, the US waits for the other country to approach it for an agreement, since it is assumed that US amateurs benefit more from such an agreement than does the other country.

Q. The Form 610-B asks for the name of a club station trustee or license custodian. What's the difference?

A. If this is a club station, then you have a club station trustee, who must have an amateur license of Technician or higher class. If this is a military recreation or RACES station then you have a license custodian, who does not have to be a licensed amateur. Only licensed amateurs may operate a military recreation or RACES station, however.

A license custodian doesn't have to have an amateur operator license because the FCC requires a military recreation station license custodian to be the person in charge of the premises where the station is located and that person is unlikely to be a licensed amateur. Similarly, the license custodian for a RACES station must be the civil defense official responsible for coordination of all civil defense activities in the area concerned [97.5(b)(3),(4)].

Q. How do we go about renewing the club license?

A. Have the station trustee complete the Form 610-B and send it to Gettysburg [97.21(a)(1)].

Q. Who is responsible for the proper operation of our club station?

A. The station trustee and the control operator are responsible for the station's operation [97.103(a)]. The control operator may operate the station only up to the privileges of his or her own operator license [97.105(b)]. If the privileges in use exceed those of the station trustee's, however, the station ID must consist of the club station's call sign followed by the home station call sign of the control op: WA1JUY/K1CE, for example [97.119(d)].

Foreign amateurs must apply for a reciprocal permit with FCC Form 610-A. (See Fig 3.) It may be obtained from the FCC or from the Regulatory Information Branch, ARRL HQ, 225 Main St, Newington, CT 06111. The *Reciprocal Permit for Alien Amateurs*, FCC Form 610-AL, is valid for one year and is not renewable, but the foreign amateur may apply for a new permit every year. A foreign amateur will have all the privileges of his or her home country as long as they do not exceed the privileges of the Amateur Extra class license of the USA. When a foreign amateur is transmitting under the authority of a reciprocal permit, he or she must include the appropriate letter-numeral designation for the US call area they are in before the call sign issued by their home country, separated by the slant sign ("/") on CW or suitable word denoting the slant sign on phone [97.119(a) and (c)].

Foreign amateurs, except representatives of foreign governments, may take the exams for FCC granted amateur licenses. In this case, they do not have to be US citizens or citizens of countries with reciprocal agreements with the US. See Table 2.

UNITED STATES OF AMERICA
FEDERAL COMMUNICATIONS COMMISSION
GETTYSBURG, PA 17326

Approved by OMB
3060-0022
Expires 11/30/92
See reverse for
public burden estimate

APPLICATION OF ALIEN AMATEUR RADIO LICENSEE
FOR PERMIT TO OPERATE IN THE UNITED STATES

ARE YOU USING THE CORRECT FORM? Use this form to request a RECIPROCAL PERMIT from the FCC to operate your amateur station in areas where the amateur service is regulated by the FCC. You must possess a valid amateur service license issued by the country of which you are a citizen. If you are a citizen of the United States, you are ineligible for a reciprocal permit—even if you are also a citizen of another country. If you will be in the United States for an extended period, you are encouraged to obtain an FCC amateur service license. If you hold an FCC-issued amateur service license, you are ineligible to be issued a reciprocal permit.

1. YOUR NAME (Last):	(First):	(MI):	2. YOUR DATE OF BIRTH: (Month, Day, Year)
3. COUNTRY ISSUING YOUR AMATEUR SERVICE LICENSE:			4. YOUR AMATEUR STATION CALL SIGN:
5. EXPIRATION DATE OF YOUR AMATEUR SERVICE LICENSE: (Month, Day, Year)			6. YOUR COUNTRY OF CITIZENSHIP:
7. YOUR UNITED STATES MAILING ADDRESS:			8. ADDRESS WHERE YOU WANT YOUR RECIPROCAL PERMIT MAILED, IF issued:
9. YOUR MAILING ADDRESS IN YOUR COUNTRY OF CITIZENSHIP:			
10. LIST ALL LOCATIONS FROM WHICH YOUR STATION WILL TRANSMIT FOR ANY PERIOD OF 30 DAYS OR MORE WHILE IN THE UNITED STATES. IF IT WILL NOT BE TRANSMITTING AT ANY ONE LOCATION DURING A PERIOD OF 30 DAYS OR MORE, WRITE "NONE":		11. APPROXIMATE DATES OF YOUR STAY IN THE UNITED STATES: (Beginning) (Ending)	

READ CAREFULLY BEFORE SIGNING

I, the above named alien amateur service licensee, request a reciprocal permit for operation of my amateur station in the United States. I understand that, if a permit is granted, the operation of my amateur station must be in accord with: (1)the terms and conditions of the agreement on this subject between my Government and the Government of the United States; (2)Part 97 of the FCC Rules; (3)the terms and conditions of the amateur service license issued to me by my Government, but not to exceed the Amateur Extra Class operator privileges, and (4)any further conditions attached to the reciprocal permit by the FCC. I also understand that any reciprocal permit issued to me may be modified, suspended, or cancelled by the FCC without advance notice. I certify that all of the information I have submitted on this application is true, complete and correct to the best of my knowledge.

WILLFUL FALSE STATEMENTS MADE ON THIS FORM OR ATTACHMENTS ARE PUNISHABLE BY FINE AND IMPRISONMENT.
U.S. CODE TITLE 18, SECTION 1001

12 YOUR SIGNATURE	13. DATE SIGNED	FOR FCC USE ONLY	EFF DATE	EXP DATE

SEE REVERSE

FCC Form 610A
January 1990

_____ **GENERAL INSTRUCTIONS** _____

A. Print clearly or type. All items must be completed, except areas indicated "FCC USE ONLY".

B. You should obtain a current copy of Part 97 of the FCC Rules from a supplier or from the Government Printing Office (GPO). For ordering information from the GPO, you may write to them at the United States Government Printing Office, Washington, DC 20402, or you may telephone them at (202) 783-3238.

C. Mail your completed application together with a photocopy of your amateur service license to the Federal Communications Commission, P.O. Box 1020, Gettysburg, PA 17326. Allow at least 60 days for processing.

D. Should you become a citizen of the United States, you are no longer eligible to obtain a reciprocal permit. If you wish to apply for an FCC amateur service license, you should contact amateur operators who are accredited as volunteer examiners. Should you obtain an FCC-issued amateur service license, it will supersede your reciprocal permit. No examination credit is allowed toward an FCC amateur service license on the basis of holding an amateur service license issued by another country.

_____ **INSTRUCTIONS FOR SPECIFIC ITEMS** _____

Item 6: State the name of the country of which you are a citizen. If different than item 3, do not submit this application. You must be a citizen of the country that issued your amateur service license to be eligible for a reciprocal permit.

Item 7: You must provide the FCC with a United States mailing address at which you may be reached or through which your mail will be forwarded during your stay in the United States. The Embassy or Consulate of your country may be able to serve this purpose for you during short visits to the United States.

Item 8: State a complete mailing address where you wish your reciprocal permit to be mailed.

Item 9: State your mailing address in your own country. If you are an alien resident in the United States, and do not have a mailing address in your own country, please state this in item 9.

BE SURE TO READ THE ENTIRE STATEMENT FOLLOWING ITEMS 10 AND 11, AND THEN SIGN AND DATE THE APPLICATION

Public burden for this collection of information is estimated to be five minutes per response, including the time for reviewing instructions, searching existing data sources, gathering and maintaining the data needed, and completing and reviewing the collection of information. Send comments regarding this burden estimate or any other aspect of this collection of information, including suggestions for reducing the burden to the Federal Communications Commission, Office of Managing Director, Washington, DC 20554, and to the Office of Management and Budget, Paperwork Reduction Project (3060-0022) Washington, DC 20503.

NOTICE TO INDIVIDUALS REQUIRED BY PRIVACY ACT OF 1974 AND THE PAPERWORK REDUCTION ACT OF 1980
Sections 301, 303, and 308 of the Communications Act of 1934, as amended, (licensing powers) authorize the FCC to request the information on this application. The purpose of the information is to determine your eligibility for a license. This information will be used by the FCC staff to evaluate the application, to determine station location, to provide information for enforcement and rulemaking proceedings and to maintain a current inventory of licensees. No license can be granted unless all information requested is provided. This information is available to the public on request. Your response is required to obtain this authorization.

Fig 3—FCC Form 610-A

Table 2

Countries that have Reciprocal Licensing Agreements with the United States

V2	Antigua/Barbuda	OH	Finland	ZL	New Zealand
LU	Argentina	F	France*	YN	Nicaragua
VK	Australia	DL	Germany	LA	Norway
OE	Austria	SV	Greece	HP	Panama
C6	Bahamas	J3	Grenada	P2	Papua New
8P	Barbados	TG	Guatemala		Guinea
ON	Belgium	8R	Guyana	ZP	Paraguay
V3	Belize	HH	Haiti	OA	Peru
CP	Bolivia	HR	Honduras	DU	Philippines
T9	Bosnia-	VS	Hong Kong	CT	Portugal
	Herzegovina	TF	Iceland	J6	St. Lucia
A2	Botswana	VU	India	J8	St. Vincent/
PY	Brazil	YB	Indonesia		Grenadines
VE	Canada	EI	Ireland	S7	Seychelles
CE	Chile	4X	Israel	9L	Sierra Leone
HK	Colombia	I	Italy	H4	Solomon Islands
TI	Costa Rica	6Y	Jamaica	ZS	South Africa
9A	Croatia	JA	Japan	EA	Spain
5B	Cyprus	JY	Jordan	PZ	Suriname
OZ	Denmark (incl.	T3	Kiribati	SM	Sweden
	Greenland)	9K	Kuwait	HB	Switzerland
HI	Dominican	EL	Liberia	HS	Thailand
	Republic	LX	Luxembourg	9Y	Trinidad/Tobago
J7	Dominica	Z3	Macedonia	T2	Tuvalu
HC	Ecuador	V7	Marshall Islands	G	United
YS	El Salvador	XE	Mexico		Kingdom**
V6	Federated States	3A	Monaco	CX	Uruguay
	of Micronesia	PA	Netherlands	YV	Venezuela
3D	Fiji	PJ	Neth. Antilles		

Notes: An automatic reciprocal agreement exists between the US and Canada, so there is no need to apply for a permit. Simply sign your US call sign followed by a slantbar and the Canadian prefix/numeral combination identifying which province you are in.

* Including French Guiana, French Polynesia, Guadeloupe, Amsterdam Island, Saint Paul Island, Crozet Island, Kerguelen Island, Martinique, New Caledonia, Reunion, St. Pierre and Miquelon, and Wallis and Futuna Islands.

** Including Ascension Island, Bermuda, British Virgin Islands, Cayman Islands, Channel Islands (including Guernsey and Jersey), Falkland Islands (including South Georgia Islands and South Sandwich Islands), Gibraltar, Isle of Man, Montserrat, St. Helena, Gough Island, Tristan Da Cuhna Island, Northern Ireland, and the Turks and Caicos Islands.

Approved by OMB
3060-0635
Expires 4/30/98
See reverse for
public burden

FEDERAL COMMUNICATIONS COMMISSION

| FOR |
| FCC |
| USE |
| ONLY |

FCC 610V - AMATEUR STATION VANITY CALL SIGN REQUEST
CAREFULLY READ THE ATTACHED INSTRUCTIONS BEFORE FILING THIS FORM

SECTION 1 - TO BE COMPLETED BY ALL APPLICANTS

1. Print or type Last Name, Suffix, First Name, Middle Initial. If a club, enter club name.

2. Date of Birth
__ - __ __ - __ __

3. Mailing Address   City   State   ZIP Code
__ __ __ __ __ - __ __ __ __

4. Payment Type Code				5. Fee Due	FOR FCC USE ONLY
W	A	V	R	$	

6. I hereby request that the currently assigned amateur station call sign _____ be vacated.

7. I hereby apply for a vanity call sign under the following eligibility (make an X in the appropriate box and enter the required information):

  7A. ☐ FORMER PRIMARY STATION HOLDER: I request call sign _____ be shown on my primary station license. This call sign was previously shown on my primary station license. **AVAILABLE UNDER ALL GATES.**

  7B. ☐ CLOSE RELATIVE OF FORMER HOLDER: I request call sign_____be shown on my primary station license. This call sign was previously shown on the primary station license of my deceased spouse, child, grandchild, stepchild, parent, grandparent, stepparent, brother, sister, stepbrother, stepsister, aunt, uncle, niece, nephew, or in-law. Enter your relationship to the deceased: _____ . **AVAILABLE UNDER ALL GATES.**

  7C. ☐ FORMER CLUB STATION HOLDER: I request call sign_____be shown on the license for the club station for which I am the license trustee. This call sign was previously shown on the license for this club station. **AVAILABLE UNDER ALL GATES.**

  7D. ☐ CLUB STATION WITH CONSENT OF CLOSE RELATIVE OF FORMER HOLDER: I request call sign _____ to be shown on the license for the club station for which I am the license trustee. The club was established prior to and was in existence on March 24, 1995. This call sign was previously shown on the primary station license of a person now deceased. I am acting with written consent of the deceased person's spouse, child, grandchild, stepchild, parent, grandparent, stepparent, brother, sister, stepbrother, stepsister, aunt, uncle, niece, nephew, or in-law. Enter the relationship to the deceased of the person giving consent:_____ **AVAILABLE UNDER GATES 1(A), 2, 3 AND 4.**

  7E. ☐ PRIMARY STATION PREFERENCE LIST: I request the first assignable call sign from my preference list in item 9 be shown on the license for my primary station. **AVAILABLE TO AMATEUR EXTRA CLASS OPERATORS UNDER GATE 2; TO ADVANCED CLASS OPERATORS UNDER GATE 3; TO GENERAL, TECHNICIAN PLUS, TECHNICIAN, AND NOVICE UNDER GATE 4.**

  7F. ☐ CLUB STATION PREFERENCE LIST: I request the first assignable call sign from my preference list in item 9 be shown on the license for the club station for which I am the license trustee.

8. Attach a photocopy of your current amateur operator/primary station license document (FCC Form 660) here. If you checked item 7C, 7D or 7F, attach a photocopy of the current club station license document (FCC Form 660) here.

FCC 610-V
November 1995

Figure 4

SECTION 2 - TO BE COMPLETED BY APPLICANTS WHO HAVE CHECKED ITEMS 7E OR 7F

9. Select your preference list of vanity call signs very carefully. Give the exact prefix, numeral and suffix for each call sign.

1. _____	6. _____	11. _____	16 _____	21 _____
2. _____	7. _____	12. _____	17 _____	22. _____
3. _____	8. _____	13. _____	18. _____	23 _____
4. _____	9. _____	14. _____	19. _____	24. _____
5. _____	10. _____	15. _____	20. _____	25 _____

SECTION 3 - TO BE COMPLETED BY ALL APPLICANTS

WILLFUL FALSE STATEMENTS MADE ON THIS FORM ARE PUNISHABLE BY FINE AND/OR IMPRISONMENT (U.S. CODE, TITLE 18, SECTION 1001), AND/OR REVOCATION OF ANY STATION LICENSE OR CONSTRUCTION PERMIT (U.S. CODE, TITLE 47, SECTION 312(A)(1)), AND/OR FORFEITURE (U.S. CODE, TITLE 47, SECTION 503).

I CERTIFY THAT ALL STATEMENTS AND ATTACHMENTS ARE TRUE, COMPLETE AND CORRECT TO THE BEST OF MY KNOWLEDGE AND BELIEF AND ARE MADE IN GOOD FAITH.

10. Signature of applicant (Do not print, type or stamp.)	11. Date signed	12. Daytime Telephone No.
X		( )

NOTE:
If none of the call signs that you requested are assignable, the call sign that you vacated will be shown on your station license.

Do **NOT** send this form to the FCC in Gettysburg, PA, Washington, DC, or any local Field Office. See attached INSTRUCTIONS for filing information.

FCC 610-V
November 1995

CHAPTER 3
Good Amateur Practice

One of the most significant, yet shortest, rules in all of Part 97 is labeled "Good amateur practice." It says, "In all respects not specifically covered by FCC rules each amateur station must be operated in accordance with good engineering and good amateur practice" [97.101(a)]. You are expected to strive to maintain your equipment, signals and operating practices in a manner consistent with the highest possible standards. This means complying with the conventions that have evolved within the amateur community to promote efficient use of our limited spectrum resources, and harmony among amateurs with different operating interests. Such conventions include band plans and other "gentlemen's agreements" not specifically addressed in the rules. Thus, it is important that you know the black-letter law of operating and the conventions that have voluntarily evolved within the amateur community, both of which comprise good amateur practice.

Peaceful Coexistence

The amateur frequencies, especially at MF (160 meters) and HF (80-10 meters) tend to be congested; interference (QRM) is a fact of Amateur Radio life. Indeed, it is a positive result of Amateur Radio growth and vitality. As such, a cardinal rule of good amateur practice is to always *listen* before you transmit on any frequency. At best, transmitting without listening is a rude interruption of someone else's conversation. At worst, it could interfere with emergency communications in progress.

If it appears that the particular frequency is unoccupied, ask "Is this frequency in use?" (QRL? on CW). If the response is a simple "Yes" (or C on CW), keep moving up the band. Once you've determined that a given frequency is clear, you're ready to operate there.

The FCC Rules clearly state that all amateur frequencies are shared and that cooperation is the key to effective use of the bands [97.101(b)]. Amateurs should subscribe to the band plans and "gentlemen's agreements" outlined in Chapter 5 as a form of this cooperation. In other cases, common sense should prevail: You don't want to ragchew with your local friend on the choicest DX portion of the 20 or 75-meter phone subband. In that case, VHF, or the high end of 20 or 75, would be a more intelligent frequency choice and consistent with good amateur practice. You are also required by the rules to use the minimum power necessary to complete the contact. Doing so helps reduce the QRM on today's crowded bands.

What about nets? No one *owns* any amateur frequency, not even a net that has been meeting on the same spot on the dial for the past 40 years [97.101(b)]. Except in an FCC-declared emergency, no one has any legal right to kick anyone else off a particular frequency, and no one's particular use of the spectrum is any more valid than anyone else's. In emergencies, however, the rules say that each control operator must give priority to stations handling emergency communications [97.101(c)].

Before starting a net, the net-control station should take precautions to avoid using a frequency occupied by other stations, by asking if the frequency is in use. Conversely, non-net operators should ask the same question to determine if a net or another QSO is in progress on that frequency. If it's a tossup, consider that many operators in a net using a single frequency is efficient spectrum usage, and it is easier for a single operator to move than it is for potentially dozens of net operators to move. Guide your operation accordingly.

The Control Operator

The control operator concept is fundamental in Amateur Radio: Some *person* must be responsible for the proper operation of an amateur transmitter. That's the control operator, defined in the rules as "an amateur operator designated by the licensee of a station to be responsible for the transmissions from that station to assure compliance with the FCC rules [97.3(a)(12)]." What that means is that the radio amateur sitting at the mike, key or keyboard, the person running the station from its *control point*, is the one who is legally responsible for the operation of the station, *regardless* of whether it's his home station or a station owned by another amateur [97.103(a)]. When the control operator of a station is someone other than the station licensee, both are equally responsible for the proper operation of the station [97.103(a)]. Unless there is documentation to the contrary, the FCC will assume that the station licensee is the control operator [97.103(b)].

Except when a station is operating under automatic control (see Chapter 6), the control operator must be present at the control point. A control operator must be at the control point when third-party communications are being transmitted, except when a station is participating as a forwarding station in a message forwarding system.

The control operator must at all times ensure the proper operation of the station, regardless of the type of control employed—local, remote or automatic [97.105(a)]. A control operator may never operate any station beyond the privileges of his or her operator license [97.105(b)]. Control of, and operation through, repeaters is discussed in Chapter 6.

The Control Point

Every amateur station must have at least one control point, that is, a location where the control operator function is performed [97.3(a)(13); 97.109(a)]. The three types of control—local, remote and automatic—are discussed more fully in Chapter 6.

Station Identification

There is a simple reason for station identification: You must clearly make known the source of your transmissions to anyone receiving them. No station may transmit unidentified communications or signals, or transmit as the station call sign any call sign not authorized to the station [97.119(a); 97.113(a)(4)].

You must identify with your call sign at the end of each contact, and every 10 minutes during the contact. You may use any emission authorized for the frequency you're using [97.119(a),(b)]. Although it's a good idea, you are not required to identify at the beginning of a QSO.

You must identify with your complete call sign, not just a portion. Some net and DX operators encourage callers to use just the "last two" letters of their call sign. This is fine as long as every operator identifies with their full call sign in accordance with the FCC requirements outlined above. If a person calls on a frequency with their "last two," they must give their full call sign at least every 10 minutes and before leaving the frequency, even if they don't make the desired contact.

Except when exchanging international third-party communications, there is no requirement to identify the other station(s) with which you are communicating. In the case of international third-party communications, at the end of the contact you must give the call sign of the station with which you exchanged third-party traffic, as well as your own call sign.

When you identify by phone, you must use English, and you are encouraged to use phonetics [97.119(b)(2)]. You can identify with a *specified* digital code when you're using RTTY for communications [97.119(b)(3)]. You can identify image (ATV and facsimile modes) using accepted color or monochrome transmission standards [97.119(b)(4)]. Spread spectrum transmissions must be identified by CW or phone [97.119(b)(5)]. When identifying with an automatic CW device, the code speed must be kept to 20 WPM or less [97.119(b)(1)].

An amateur station must identify itself on every channel on which it is transmitting. Repeaters, remote bases or crossband radios must be properly identified on all outputs. If you are communicating through a repeater, there is no need to state that fact, or identify the repeater; the repeater is required to identify itself in accordance with the rules.

On the Road

While most of your ham radio activity will probably take place at your home station, you may operate from other locations—your car, your vacation spot, at Field Day or even outside the country. Amateurs are permitted to operate their stations portable or mobile anywhere in the US, its territories and possessions. There is no need to notify the FCC, or keep a log of your portable operation in these areas (note that when there is a *permanent* change in your mailing address, you must notify the FCC on Form 610). You don't have to sign "mobile" or "portable." If you're operating your station away from home for a long time, arrange to have your mail forwarded to your temporary location, in the unlikely event that the FCC sends a Notice of Violation or other official correspondence to your permanent mailing address.

Drawing a distinction between mobile and portable operation is sometimes helpful. Mobile refers to talking on your 2-meter FM rig while driving your pickup truck, or while jogging or backpacking. Portable means operating for an extended period at a specific, definable location, such as your cabin in the mountains or condo in Florida.

The ARRL DXCC desk recognizes Puerto Rico as a separate country, but the

Q&A—Station Identification

Q. Do I have to give my call sign at the beginning of the contact?

A. You only have to ID at the *end* of the QSO and at least once every 10 minutes during its course [97.119(a)]. There is no requirement to transmit your call sign at the beginning of a contact.

Q. How often do I have to give the call sign of the station I'm talking to?

A. You are not legally required to mention the other guy's call at all [97.119(a)]. The only exception is when handling traffic with foreign stations—you must then give the other station's call sign at the end of an exchange of third-party traffic [97.115(c)].

Q. When my ham friends from the local radio club come to visit my home and operate my rig, or vice versa, whose call sign is used? What frequency privileges may they (I) use?

A. Every control operator is bound by the frequency privileges of his or her license, regardless of what class of license is held by the station owner [97.105(b)]. For ID purposes, let's use an example: If Joanne, KA1SIP, a Technician, visits the shack of Steve, WV1X, an Amateur Extra, Joanne uses Steve's call sign, but must stay within her Technician privileges. If Steve authorizes it, she may use her own call at his station, but must stay within her Technician privileges. She could also operate as a third party (that is, as an unlicensed individual) outside her Technician privileges as long as Steve acts as the control operator and remains at the control point continuously monitoring and supervising the operation [97.115(b)]. On the other side of the coin, if Steve visits Joanne's shack and is using his Extra Class privileges, Steve must add his own call sign to the end of the station ID: KA1SIP/WV1X [97.119(d)]. Joanne could also "lend" Steve her station, and then Steve operates his "temporary" station under his own station/operator license and call sign—no special ID needed.

Q. I regularly conduct my QSOs in Esperanto. Is it okay to identify in Esperanto?

A. Sorry, Esperanto doesn't cut it. When operating phone, you must identify your station in the *English* language [97.119(b)(2)].

Q. What's this "Temporary AG" stuff I've heard on the HF bands?

A. If you have a license and pass an upgrade examination, the Volunteer Examiners will issue a Certificate of Successful Completion of Examination (CSCE), a temporary permit that allows the use of your new privileges immediately [97.9(b)]. In the interim period after passing the exam but before the FCC grants the upgrade, you must add a special designator to your call sign whenever you use your new privileges. For example, if you just upgraded from Novice to General and you are operating in the General subband, you must give your call sign followed by the words "Temporary AG," or the slant mark (/) followed by the letters AG on CW. The AG is the FCC's indicator for General class (the other indicators are KT for Technician, AA for Advanced and AE for Extra Class) [97.119(e)].

Q. This summer, G5BFU from England and VE3GRO from Ontario are coming to visit me and my family. How should G5BFU and VE3GRO identify while they're here in the States?

A. Let's say G5BFU gets his permit and is on the air from sunny California; his station ID consists of the US letter prefix and number followed by his home call sign: "W6/G5BFU." This is the European style of portable identification that the FCC adopted in June of 1988. Because of treaty constraints, however, proper identification for VE3GRO is in the traditional manner—Canadian call sign *followed by* US prefix. VE3GRO in 6-land would ID as "VE3GRO/W6." The "W6" portion of the identification is considered the "indicator" by the FCC. The call sign and the indicator are separated by the fraction bar (/) on CW or RTTY, or a pause (that refreshes) or the words "stroke," "slant," "slash" or "portable." The rule also requires that at least once during the QSO, the city and state must be mentioned, in English. G5BFU, who we wouldn't expect to have much trouble with the English language might ID: "This is W6 stroke G5BFU in beautiful downtown Burbank, California" [97.119(f)].

Q. Is it okay to give my call sign in phonetics when identifying?

A. Yes. The FCC encourages the use of a nationally or internationally recognized standard phonetic alphabet *when necessary* as good amateur practice [97.119(b)(2)]. The International Telecommunication Union (ITU), an agency of the United Nations, has developed such a list, and the League encourages its use when phonetics are necessary to make the *initial* identification (such as when signals are weak, or in a DX pileup when there is QRM). The ITU list is as follows

A-Alfa	B-Bravo	C-Charlie	D-Delta
E-Echo	F-Foxtrot	G-Golf	H-Hotel
I-India	J-Juliette	K-Kilo	L-Lima
M-Mike	N-November	O-Oscar	P-Papa
Q-Quebec	R-Romeo	S-Sierra	T-Tango
U-Uniform	V-Victor	W-Whiskey	X-X-Ray
Y-Yankee	Z-Zulu		

For example, Shelly, WB1ENT, may phonetically identify on phone by giving her call sign as "Whiskey Bravo One Echo November Tango."

Q&A—Operation Away from Home

Q. My wife and I are setting sail on a luxury liner for a cruise among the Caribbean islands. I'd like to bring my ham rig with me. When I operate, whose rules do I follow?

A. The first determinant is the country of registry for the vessel. When you operate on the high seas (i.e., international waters) on a US-registered vessel, you must follow the FCC Rules. US and Canadian licensees need no special permit or authorization other than their own amateur licenses for operation aboard a US-registered vessel; these two countries share an automatic reciprocal-operating agreement [97.5(d)(6)].

If you are an alien (other than a Canadian citizen) licensee, you must obtain a US amateur license or a reciprocal operating permit from the FCC *prior* to your operation aboard the US ship. A reciprocal permit may be issued by the FCC to licensed amateurs of a foreign country that has signed an agreement with the US to issue such permits. The amateur applying for such a reciprocal permit must be a citizen of that country (his home country) [97.5(d)(5)].

Q. Whose rules do I follow when my US ship sails into the territorial waters of another country?

A. When sailing or anchored in the territorial waters of another country, you must check the radio rules of that country before you operate your station. You must comply with those rules and obtain any required license or permit from that country's government. It is recommended that you study the country's requirements in advance of your departure date (send a 9 × 12-inch SASE with one unit of postage for each country requested to the Regulatory Information Branch at ARRL HQ for information), as some administrations can take as long as six months or so to issue the necessary operating paperwork. Plan ahead! Never operate in a foreign country or in its territorial waters without proper authorization.

Q. What if the ship is of non-US origin, even though I'm an FCC-licensed amateur?

A. If you are sailing in international waters aboard a non-US vessel, check the amateur rules of the country of the ship's registry (contact the Regulatory Information Branch at HQ for information on a particular country). Because a ship is considered a part of the country in which it is registered, you must obey those rules and obtain any necessary license or permit from that country's government prior to your operation. If you are sailing in the waters of a foreign government's territory, you must observe the rules of that country and obtain any necessary license or permit from that government prior to your operation in its territory. You have a two-fold responsibility: To obey the rules of the country of the ship's registry *and* the rules of the country(ies) whose territorial waters you're operating from.

Q. What do foreign Amateur Radio rules involve and how are they different from Part 97?

A. Generally, they are the same subjects as covered by our Part 97, although they tend to be more concise. Frequency privileges, power limitations, ID requirements, third-party traffic restrictions and other regulations vary widely from country to country. Hams in the US often assume that Amateur Radio operators in other countries have been allocated the same frequencies as US hams. This is often not the case. Moreover, each of the three International Telecommunication Union (ITU) Regions has its own allocations for the various services, and they do not necessarily conform to our frequencies in the US (Region 2). When operating in a foreign country and/or on a foreign vessel, always observe *all* of their rules.

Q. What US rules apply when I'm flying or sailing over/in US territory?

A. Concerning your equipment aboard a ship or plane, the rules require that the installation and operation of your station must be approved by the master of the vessel or the pilot [97.11(a)]. You must keep your radio gear separate from and independent of all radio equipment installed on the ship or plane, except a common antenna may be shared with a voluntary ship radio installation [97.11(b)].

Your transmissions must not cause interference to any equipment installed on the ship or plane [97.11(b)]. Your station must not pose a hazard to the safety of life and property [97.11(c)].

You must not operate your rig on a plane when it is flying under Instrument Flight Rules (IFR), unless you can be sure your station complies with all applicable FAA Rules [97.11(c)]. When you operate in ITU Region 2, frequencies used must be consistent with US frequency bands and if you are a US licensee, you must operate within the privileges of your license. Outside Region 2 and subject to the limitations of your license class, other frequency segments may apply [97.301]. For example, on 40 meters in Region 2, 7.0-7.3 MHz is allocated to the Amateur Radio Service, but in Regions 1 and 3, the amateur allocation is only 7.0-7.1 MHz.

Q. What ITU Region am I in?

A. If you're operating from Europe or Africa or the adjoining waters, you are in ITU Region 1. North and South America and the adjoining waters comprise Region 2. The rest of the world (the countries of Southern Asia and the islands of the Pacific and Indian Oceans) make up Region 3. See the map in Chapter 5.

Q. What type of station ID procedure should I use when I operate maritime mobile in international waters?

A. FCC Rules place no special identification requirements on a station in maritime (or aeronautical) mobile operation. It is helpful to listeners, however, if you follow your call sign with the words

"Maritime (or aeronautical) Mobile" followed by the ITU Region in which you are operating. You can send your call sign followed by the fraction bar, then the indicator MM (for maritime mobile) and the number of your ITU Region, such as KB1MW/MM2. Your ITU Region will always be 1, 2 or 3 (don't confuse ITU regions with domestic US call areas).

Q. If I'm sailing in international waters, can I pass third-party message traffic?

A. In US territory or on a US-registered vessel in international waters, you may handle third-party message traffic, including phone patches, to countries with which the US holds third-party traffic agreements [97.115]. Certain countries, however, *prohibit* third-party traffic handling, or in some cases, prohibit phone patches, while permitting other forms of message handling. You can't pass message traffic with countries that have *not* signed a third-party traffic agreement with the US. The only exceptions are when the third party is a licensed amateur, and/or in a *bona fide* emergency. See Chapter 7 of this book for a complete discussion of third-party traffic handling, including the ARRL phone patch and autopatch guidelines.

Q. How far do the territorial waters of a particular country extend?

A. The territorial limits extend to wherever that particular country says they do. The US, Canada and Mexico claim 12 nautical miles. Most Caribbean island nations claim 3-12 miles, although there are exceptions. The territorial limits of certain Caribbean nations and other island nations whose area consists of many small islands often overlap, creating a larger territorial area than one would assume.

Q. A ham friend of mine told me that FCC-licensed amateurs within international waters could operate on all frequencies, regardless of their license class. Is this true?

A. No way. If an FCC-licensed station operates outside the privileges of his/her license class, it's a violation of FCC Rules [97.105(b)].

FCC doesn't. US amateurs sometimes forget that when they go to American Samoa, the US Virgin Islands or any other place where Amateur Radio is regulated by the FCC, they may operate normally under the terms of their license. See p. 97-48 for the places where the Amateur Service is regulated by the FCC. When operating from US territory that has a distinctive prefix such as KH6 (Hawaii), you are not required to sign with that prefix, but you may do so to let others know you're "DX."

Operating in Other Countries

An FCC license permits you to operate only in the United States and its territories. The US has entered into reciprocal operating agreements with a number of foreign countries; there is a list of them in Chapter 2.

Operating at Sea or in the Air

When you are operating aboard a ship or an aircraft, the FCC says your amateur station must be separate from all other radio apparatus, except that a common antenna may be used. You must also have permission of the master or pilot in command [97.11].

Gray areas sometimes occur in maritime mobile operation in and around foreign ports. Well-intentioned US hams often have trouble determining whose rules they're supposed to follow when sailing the seven seas. Certain people, regrettably, are not so well intentioned. Unscrupulous yachters and smugglers without ham tickets have been known to invade amateur frequencies for purposes such as ordering parts and supplies for their vessels, conducting stateside business affairs, or worse. They cling to the mistaken notion that because they're on the high seas, the rules do not apply to them. Others forego standard marine-mobile emergency communications gear, relying instead on an unlicensed ham transceiver—a dangerous proposition for an untrained sailor in a maritime emergency.

When you are on board a vessel in US territorial waters, you may operate under the terms of your FCC license. When in the territorial waters of another country, you must have a license or reciprocal operating permit from that country. Territorial waters usually extend from 3 to 12 (or more) nautical miles off shore. The United States claims a 12 nautical mile territorial limit. When you are beyond any country's limits you are in international waters; a vessel in international waters is effectively a part of its country of registry. If it's a US vessel, then you may operate under your FCC license. On the other hand, if the vessel is registered in another country, you must get a permit or license from that country.

Sometimes even licensed hams disregard the rules, complicating US and foreign efforts to reach important third party and reciprocal operating agreements. Administrations unfriendly to international Amateur Radio are the first to cite these instances when attempting to remove privileges and frequencies at worldwide allocation conferences.

Conclusion

The four cornerstones of successful ham radio operating are: knowledge, skill, dedication and courtesy. This chapter is intended to explain the basic rules for operating an amateur station. Skill is a product of operating ability and repeated practice so the mechanics become almost second nature. Dedication involves spending the time to learn, then applying that to your enjoyment of this rich hobby. Courtesy in Amateur Radio goes beyond just being considerate and extends to being helpful to others.

Good amateur practice in on-the-air operating means all of the above, that is, being guided by the intent *and* the spirit of the rules. The FCC and the amateur community have worked together to forge a consensus represented in Part 97 that makes Amateur Radio operating dynamic and creative.

CHAPTER 4
Authorized and Prohibited Practices

The rules are generally set up to allow flexibility and breathing room. You can experiment with new ideas, help the public enjoy and participate safely in events such as parades and walkathons by providing support communications, and promote goodwill between us and the peoples of other nations. The rules allow for providing emergency communications in support of disaster relief efforts, a major reason for the existence of Amateur Radio.

In this chapter we'll examine some main points that Amateur Radio operators can and cannot do. Other chapters explore how stations may be operated, and what frequencies and emissions can be used to get the most out of our service. There are, however, a few specific things you must *not* do—the FCC calls them "prohibited transmissions" [97.113]. Many of these are common sense, black-and-white, needing little explanation or interpretation. But others lie in the gray area between black and white. These require interpretation and clarification for specific situations. We will also look at specifically authorized transmissions.

Because it is usually dangerous for the lay ham to ask FCC officials for interpretations, we will present comprehensive discussions of FCC letters, news releases, public notices, approved interpretations and so forth that will help you understand the gray area rules and apply them to your activity. We hope you will not need to consult the FCC after you read these interpretations.

The rules we will be looking at are important. Some are required to preserve the core of Amateur Radio: the prohibitions against broadcasting to the public, and operating your station for compensation and on behalf of your employer, for example.

Authorized Communications

Amateur Radio is a two-way radio service. Communicating with other amateur stations is what it's all about. The exception is with stations in countries whose administrations have banned such communications. At present, there are no "banned" countries, but the FCC will tell us in a public notice if this changes [97.111(a)(1)].

Speaking of international communications, use plain language, and confine your conversations to technical and personal comments relating to equipment tests, idle chatter and so forth when working stations in other countries. The content of your conversation must be unimportant to the point where no party would be compelled to use the public telecommunications system to communicate the same information [97.117]. This protects foreign governments' interests in public telecommunications revenue.

You can communicate with stations in other FCC-licensed radio services during emergencies and disasters. For example, from time to time, Military Affiliate Radio System (MARS) and Coast Guard stations have come into the amateur bands to work directly with amateurs to save lives in emergencies [97.111(a)(2)]. Further, you can communicate with US Government stations when providing communications in the Radio Amateur Civil Emergency Service (RACES) [97.111(a)(3)]. We will discuss RACES later in this book.

You may also communicate with radio services not regulated by the FCC, provided that they are authorized by the FCC to communicate with Amateur Radio stations. Hams also pass messages with military stations during Armed Forces Day communications tests [97.111(a)(4)].

Because Amateur Radio is primarily a two-way communications service, there are only a few cases when you can send a one-way transmission. You can send short transmissions for:

- testing your equipment (for the "minimum time necessary"). Note that good amateur practice dictates the use of a dummy load and not tuning up on a pileup on a DX station. [97.111(b)(1)]
- calling CQ [97.111(b)(2)]
- sending telemetry or telecommands to remote stations [97.111(b)(3),(7)]
- emergency transmissions [97.111(b)(4)]
- information bulletins of interest to amateurs only, and code practice, such as that transmitted by ARRL Headquarters station W1AW [97.111(b)(5),(6)].

Direct and Indirect Payment

You must never accept any money or other consideration for operating your station [97.113(a)(2)]. This is consistent with one of the prime directives of our service:

"Recognition and enhancement of the value of the amateur service to the public as a *voluntary, noncommercial* communication service, particularly with respect to providing emergency communications" (emphasis added) [97.1(a)].

If your club is providing communications support to the town of Needham for a parade, you cannot accept their offer of payment for your work. You are volunteers, providing a community service on a noncommercial basis, period.

You should never accept anything for your Amateur Radio operating. The FCC prohibits operation of an amateur station "for hire, or for material compensation, direct or indirect, paid or promised" [97.113(a)(2)]. This includes direct payment (money, goods, food, and so on) and indirect payment (publicity, advertising, and so on).

Back to our example, the Needham town officials may, however, provide you with items to verify the identity of participating amateurs. Badges, caps, T-shirts, signs and other incidental items are not considered to be material compensation [ARRL opinion].

How is it that W1AW (the ARRL HQ station) operators are allowed to be paid? The rules provide for such an activity. Control operators of a club station may be compensated when the club station is operated primarily for code-practice transmissions or for transmitting bulletins consisting of Amateur Radio news having direct interest to hams. To qualify under this provision, however, the station must conduct code-practice sessions for at least 40 hours per week, schedule operations on at least six medium- and high-frequency amateur bands, and publish the schedule of operating times and frequencies at least 30 days in advance of the actual transmissions. Control operators may accept compensation only when the station is transmitting code practice and bulletins [97.113(d)].

Business Communications

From 1972 to 1993, the FCC laid down stringent "no business" rules. Effective September 13, 1993, the "no business communications" language was replaced with a prohibition on communications for compensation, on behalf of one's employer, or in which the amateur has a pecuniary interest [97.113(a)(2),(3)]. The current language is almost, but not quite, as relaxed as the pre-1972 rules. Now, instead of a flat prohibition on providing an alternative to other radio services, there is a less restrictive one against doing so on a regular basis [97.113(a)(5)]. Teachers are specifically permitted to make incidental use of Amateur Radio in their classroom instruction without having to worry about the "no compensation" rule [97.113(c)].

The new rules permit wider use of Amateur Radio to satisfy personal communications needs. To cite a classic example, as far as the FCC is concerned you can now use an autopatch to order a pizza. You can call your dentist's office to let them know you'll be late, or even to make an appointment. On your way home you can ask your spouse if you should pick up a loaf of bread without worrying about whether this will "facilitate the business affairs" of the grocery store. Repeater owners or trustees can set tighter standards if they want, but it's no longer an FCC issue.

The Commission doesn't want to hear questions about whether such-and-such is permitted. The Report and Order said, in part, that " . . .any amateur-to-amateur communication is permitted unless specifically prohibited, or unless transmitted for compensation, or unless done for the pecuniary benefit of the station control operator or his or her employer" [PR Docket 92-136, Report and Order].

A simple check list may help you determine if a communication is permissible under 97.113:

1) Is it expressly prohibited in the rules (music, obscenity, etc.)?
2) Is it transmitted for compensation?
3) Does the control operator have a pecuniary interest, i.e., could he benefit financially?
4) Does the control operator's employer have a pecuniary interest?

If you can answer "No" to all of these questions, the communication is okay as far as the FCC is concerned.

Broadcasting

Broadcasting refers to transmissions intended to be received by the general public. As such, broadcasting must be left to those services directly authorized to do so by the FCC—commercial television and radio, for example. Broadcasting is specifically prohibited in the Amateur Radio Service [97.113(b)].

You can't transmit anything intended to be received by the public directly or by intermediary relay stations. It's important to note the difference, however, between broadcasting and the permitted one-way transmissions of bulletins. Bulletins dealing directly with Amateur Radio matters may be transmitted one-way to amateurs only; they may not be transmitted for the benefit of the general public. For example, an amateur may recite the latest ARRL bulletin over his local repeater, but may not give a rundown on the current Middle East situation intended for the ears of the general public with home scanners. You may not retransmit AM or FM broadcast programs or television audio or video. You may, however, retransmit NOAA weather broadcasts, propagation bulletins originated by US Government stations, and space shuttle communications (with NASA's approval) on an occasional basis, as an incident to normal amateur activity [97.113(e].

Occasionally, at a Field Day or other operating activity, a local broadcast station may ask you to let them record your station's transmissions and receptions for the local evening news. This is fine. You may consent to the retransmission, live or delayed, of your amateur station transmissions by a radio or television broadcast station.

One-Way Information Bulletins

The best-known one-way bulletins come from W1AW. For decades, W1AW has transmitted brief information bulletins. It is common practice for other amateur stations to conduct similar operations, usually on a localized basis, and as part of a scheduled net or on some other limited, scheduled basis. Such bulletins are important to the dissemination of timely and accurate information to radio amateurs.

W1AW and other stations providing this service generally have been careful to make such bulletins brief and factual. Occasionally, such as during the launch of an Amateur Radio satellite or space shuttle, some amateur stations using phone have been put into a quasi-broadcast operation to provide real-time information about a developing situation of special interest to radio amateurs. Such events are rare and amateurs have exercised restraint [97.113(b),(e)].

According to the rules, the content of bulletins must be limited to information on matters of direct interest to radio amateurs only, and never intended for the ears of the general public [97.3(a)(24)]. As with business communications, the potential for abuse of the one-way transmissions rule has long existed. There is no room in the crowded amateur high-frequency bands for lengthy one-way transmissions on a regular, quasi-broadcast basis. Information bulletins by definition must be confined to factual matters, not matters of opinion. They must be intended to inform, rather than to entertain. [97.111(b)(6)].

Newsgathering and the Media

You must not allow your amateur station to be used for any activity directly related to program production or newsgathering for broadcast purposes [97.113(b)]. There is only one exception. You can provide news information about an event if the following requirements are met: The information involves safety of life and/or property; is directly related to the event; can't be sent by any other means because normal systems are disrupted or aren't available at the site where the information is originated; and other communications could not be reasonably provided before or during the event. These rules protect the amateur service from encroachment by commercial news media that would use Amateur Radio as an inexpensive alternative to its more expensive systems [97.113(b)].

Music

The transmission of music is strictly prohibited in the amateur bands [97.113(a)(4)]. For example, expect some mail from the FCC if you transmit Rossini's *The Barber of Seville* on your favorite 75-meter phone net. The FCC mail will not be a critical review of the performance; rather, it may be a Notice of Apparent Liability requiring a formal reply within 10 days of receipt.

The FCC has specifically stated that incidental music on space shuttle retransmissions is okay [97.113(e)].

Criminal Activities

Amateur Radio may not be used for any purpose or in connection with any activity contrary to federal, state or local law. This means that if you use your 2-meter hand-held to communicate to your getaway car during a bank heist, you'll not only be in trouble with the cops, but with the FCC as well. On top of getting 15 to 20, you could lose your license! [97.113(a)(4)].

Codes and Ciphers

You may not hide the meaning of your communications by putting them into codes or ciphers [97.113(a)(4)]. In this context, codes and ciphers refer to text that has been transformed to conceal its meaning. This restriction is consistent with the Commission's obligation, under international regulations, to ensure that amateur

Q&A—Music

Q. At last night's club meeting, a fellow member told me that he could hear my car radio playing in the background during our repeater QSO. I didn't even think about it at the time—but, was I in violation of the rules?

A. What the FCC is really concerned about is the deliberate playing of music on Amateur Radio for purposes of providing entertainment. The FCC doesn't want the general public tuning in to your amateur station to listen to music!

As a practical matter, you won't go to jail for an isolated incident if you're talking on 2 meters. But, if the background radio music was loud enough to cause your friend to mention it to you, it was too loud. It might have disturbed others, too. Just keep it down next time!

Q. My friend and I are music composers. Can we trade compositions in ASCII code across town on 2 meters?

A. The music prohibition concerns the playing of music itself, not information about music. As long as no musical notes can be detected on the air, you're okay.

Q. What if we send commands to each other's MIDI synthesizers over the air?

A. Again, no problem, as long as no actual music is played on the air and the transmission took place on 50 MHz or higher [97.307(f)(5)].

Q. Can we sing "Happy Birthday" to our friend on the local 2-meter repeater?

A. No. Singing is music, and is prohibited, no matter how badly you sing.

communications be made in plain language. Universally accepted abbreviations may be used, however, when the intention is not to hide the meaning of the transmission; common Q signals, for example.

The rules do not apply to the legitimate use of secret telecommand codes used to protect the security of a space station [97.207(f)].

Also, repeater trustees can use authentication codes to secure access to the control functions of their hilltop repeater. (Authentication codes are sent by control operators to control repeater functions—the repeater checks, or authenticates, the code to ensure that the commands are coming from an authorized control operator.) Universally accepted touch-tone pad tones are not considered to be codes and ciphers; their use in traditional repeater and other applications is fine.

Obscenity and Indecency

The rules say you cannot transmit obscene or indecent words, language or meaning [97.113(a)(4)]. But when is the line crossed? In April 1987, the FCC

FCC Cites Amateur for Indecent Speech

On September 21, 1992, the Commission fined a Tennessee amateur $2000 for violation of §97.113(a)(4). The notice stated, "On June 29, 1992, between 3:53 and 4:22 PM CST, the Commission's Kingsville Office monitored a conversation on frequency 14300 kHz which included transmissions by an individual identifying with the call sign. . . We find that the words and language transmitted are indecent within the meaning of Section 97.113(a)(4) and prevailing Supreme Court and Commission precedent." The release went on to state, "Speech is considered indecent if it describes sexual or excretory activities and organs in patently offensive terms. It need not depict or describe 'hard core' sexual conduct as required for a finding of obscenity. We find that the transmissions describe sexual acts and organs in a patently offensive manner and go well beyond what an average adult person in any community would consider to be worthy of protection."

The Commission stated further, "One of the Commission's goals is to protect children from exposure to sexually explicit communications over the airwaves. The Commission has previously stated that 'the concept of indecency is intimately connected with the exposure of children to material that most parents regard as inappropriate for them to hear.' Amateur transmissions are accessible to homes over a large area and a significant number of licensed amateurs are children in their formative years. Additionally, the transmissions were made in the summertime in the afternoon when there is a real likelihood that children are listening."

The Commission's action signals the amateur community that the agency will continue to enforce its rules in this arena.

notified licensees of standards that would be applied in these cases:

Obscenity: (1) An average person, applying contemporary community standards, must find that the material, as a whole, appeals to prurient interest; (2) The material must depict or describe, in a patently offensive way, sexual conduct specifically defined by the applicable state law; and (3) The material, taken as a whole, must lack serious literary, artistic, political, or scientific value.

Indecency: Language or material that depicts or describes, in terms patently offensive as measured by contemporary community standards for the broadcast medium, sexual or excretory activities or organs.

Is the use of an expletive in itself indecent? According to the Commission, "deliberate and repetitive use of such expletives in a patently offensive manner would be a requisite to a finding of indecency." The context is also important.

False Signals

False or deceptive messages, signals and identification are clearly prohibited in the amateur bands [97.113(a)(4)]. You may not use someone else's call sign without authorization, or transmit a communication intended to deceive listeners. Don't retransmit other amateurs' communications without their knowledge or without properly identifying such retransmissions. Sending a false "SOS" or "MAYDAY" won't win you any friends, either [97.113(a)(4)].

Retransmitting Radio Signals

Radio amateurs may only retransmit signals originating in the Amateur Radio Service, except in cases of emergency, or communications on US Government frequencies between a space shuttle and its associated Earth stations where NASA permission has been granted. In addition, amateurs may transmit NOAA weather and government propagation bulletins. Such retransmissions must be limited to an occasional basis [97.113(e)]. Some amateurs find that retransmitting a friend's 20-meter signal over the local repeater adds variety to their day-to-day activities. No problem, as long as this is done manually (with control operator present). You must never record someone else's transmissions and play them back over the air without permission. That's poor amateur practice [97.101(a)] and transmission of false signals [97.113(a)(4)].

Automatic retransmission of amateur signals may be done only by space, repeater or auxiliary stations [97.113(f)]. Message forwarding stations participating in a message forwarding system may be automatically controlled when transmitting third party communications [97.109(d),(e)].

Malicious Interference

A combination of FCC-mandated and voluntary restrictions intended to keep us out of one another's way help, but these cannot completely eliminate QRM (interference) between amateur stations—nor should we expect them to.

Indeed, zero QRM is an utterly unrealistic expectation in Amateur Radio. If interference-free communication is your primary goal, you're in the wrong hobby—you should be experimenting with fiber optics. We will attempt to put QRM into perspective. Note that we're referring only to interference from one amateur station to another, not RFI/TVI or to nonamateur intruders into exclusive ham bands.

Except when it concerns emergency communications, amateur-to-amateur interference is not, in and of itself, illegal. Each amateur station has an equal right to operate; just because you've used the same frequency since 1947 doesn't mean you have any more legal right to it than the guy who got his license in the mail five minutes ago. The rules specifically prohibit willful or malicious interference [97.101(d)].

What's malicious interference? Here's an example. If two hams, or groups of hams, find themselves on the same frequency pursuing mutually exclusive objectives, that's happenstance, not malicious interference. On the other hand, if one moves to another frequency and the other follows for the purpose of continuing to cause QRM to the first, the second has crossed the line. If he does it enough, he'll put his license in jeopardy. Of course, what sometimes happens is that they'll all sit on one frequency and argue about who has more right to be there. All it accomplishes is to keep the frequency from being used by anyone for anything worthwhile.

Other FCC Rules are intended to minimize (but not to eliminate) interference. One says, ". . . amateur stations must use no more than the minimum transmitter power necessary to carry out the desired communications" [97.313(a)]. There's also a requirement that ". . . each amateur station shall be operated in accordance with good engineering and good amateur practice" [97.101(a)]. But hams' obligations to each other don't end with the rules; even more important is the need for common sense and common courtesy, when hams share common spectrum resources.

Radio amateurs have the right to pursue legitimate objectives within the privi-

FCC Actions Help Define Malicious Interference

Two recent enforcement actions by the Commission will serve to help amateurs when it comes to defining what is actionable under the prohibitions against malicious interference.

The first, adopted on November 3, 1992, was a fine of $2000 against an amateur for willfully and maliciously causing interference to other amateur operations. The FCC said that the amateur "deliberately used [his] signals to interfere with the communications of other amateur operators on 7257 and 7258 kHz." Further, "these transmissions continued throughout the period 9:03 AM to 9:52 AM. [The operator] willfully operated on frequencies he knew were being used and harassed other operators with rude and insulting communications." The operator identified himself during the communications. "When other operators tried to switch frequencies, [the operator] followed them onto the new frequency and continued to verbally harass them," the FCC said.

In a January 19, 1993 release, the FCC fined another amateur licensee $10,500 for violation of the rule. On September 26, 1992, from 7:54 to 8:27 AM CST, the FCC Kingsville Office monitored and recorded Amateur Radio communications on 14.314.7 kHz, including willful and malicious interference by the subject. The Kingsville office subsequently sent a violation notice to the subject. A transcript of the communications giving rise to the violation accompanied the notice.

The subject replied and did not dispute making the transmissions, but denied that his signals constituted willful, malicious interference, because, he said, if he truly wanted to willfully and maliciously interfere he could have done so to such an extent that communications between other Amateur Radio operators would have been impossible; that his comments were not disruptive; that his statements were innocuous; that he was recognized and spoken to by the other amateur operators; and that he was the victim of entrapment by the other operators. The FCC dismissed the subject's answer and said that the fact remains that he willfully violated §97.101(d) and that forfeiture penalties are warranted. The FCC also said that the subject had violated the rule in the past and that this had been brought to his attention by two previous notices. The fine was adjusted upwards accordingly.

leges conveyed by their licenses; but they also have the obligation to minimize the inconvenience and loss of enjoyment hams cause to others. If there's a tiny segment of a band used for international communication, it's not too much to ask that local ragchews take place elsewhere. If establishing a beacon in the middle of a densely populated area is going to QRM nearby weak-signal enthusiasts, an amateur can find another place to put it. And surely, in such cases amateurs don't need the FCC to tell us what growing up in a civilized society should already have taught us to do.

Swap Nets

The legality of nets where members make arrangements to buy, sell and trade Amateur Radio gear has traditionally been one of the great debates. Fortunately, the FCC has dealt with the issue by putting its policy into the rules. You are allowed to inform other amateurs of the availability for sale or trade of ham gear, unless you are trying to make a profit by buying or selling gear on a regular basis. Thus, under these conditions, Swap Nets are okay [97.113(a)(3)]. The FCC staff has also said that mentioning the price over the air is okay, but the "haggling" should be handled on the telephone.

Summary: "Is It Legal?"

Next time you're wondering whether something's legal, first check for the letter of the rules. If you can't find a specific rule, check the spirit of the rules found in many places, but never more prominently than in the Basis and Purpose of Amateur Radio. Use this *Rule Book* to help you decide. We've included many interpretations that have evolved over the years to help you apply the rules to your

Q&A—Malicious Interference

Q. How can I fight malicious interference on a personal level?

A. When you experience malicious interference, use the most effective means of combating it at your disposal—ignore it on the air. Above all, keep your on-the-air operation strictly above reproach. Don't engage interfering stations in their own game. By ignoring the offenders, you deprive them of their number one need—attention.

Q. What can we do about a jammer on our repeater?

A. The FCC Amateur Auxiliary incorporates a special function in its program to deal with repeater problems—the Local Interference Committee (LIC). The LIC is organized at the local level to track down repeater jammers and other problem-causers, to exert peer pressure to clear up problems, and when problems persist, to work to gather evidence leading to an enforcement action. The LIC is officially sanctioned by the ARRL Section Manager. If you've got a jammer on your hands, contact your Section Manager.

Q. Does the system for shutting down jammers work

A. Yes. Here's what happened in a recent case in Washington state: FCC News Release—"Working closely with the Amateur Auxiliary, the Federal Communications Commission (FCC) was able to issue a Notice of Apparent Liability...

The action against the amateur was triggered by numerous complaints from other Amateur Radio operators that [the amateur] was transmitting on 146.04 MHz for the sole purpose of disrupting the communications already in progress on that frequency. The intentional jamming signals were directly observed and identified by members of the Amateur Auxiliary's Local Interference Committee (LIC). The LIC forwarded the pertinent information to the local FCC office for investigation and enforcement."

Q. Our public service net has been plagued with a ham who

situation. The plain-language discussions should also help. If you still can't decide, contact ARRL HQ's Regulatory Information Branch for help. Its staff is trained in helping amateurs understand and interpret the rules. Contact them at:

ARRL Headquarters
Regulatory Information Branch
225 Main St
Newington, CT 06111

Don't call the FCC! Usually, amateurs contacting the FCC do so for emotional reasons. Harmful precedents can be set by an FCC official answering a poorly crafted question. The cure can be far worse than the temporary affliction.

---

harasses us—it seems he'll do anything to disrupt our good work. To make matters worse, others join in with catcalls, whistles and other attempts to get rid of the guy. Is this malicious interference, and what can we do about it?

A. Nets are sometimes singled out for abuse because of their "ownership" of a particular frequency. Of course, no individual operator or group has any special privilege to any part of a band.

Although you are suffering a considerable amount of thoughtless remarks, annoying practices and harassment, there seems to be no evidence of malicious interference. To be actionable, interference must be truly malicious (see above). To nail an amateur on a malicious interference charge, Amateur Auxiliary members collect evidence in accordance with the comprehensive guidelines outlined in Chapter 16. The evidence must include dates, times, frequencies, call signs, local field-strength readings and direction-findings of the offending transmissions so that re-creations of factual occurrences can be made. Conclusions must not be drawn by the monitors—it is for the FCC to determine whether the evidence, taken together, shows that a rule has been violated and that the accused person did it.

The task before the Auxiliary in a case such as yours is not an easy one: There are numerous stations responding to the primary offending operator with their own jamming transmissions, aggravating the situation enormously. The good news is that evidence gathered in accordance with the guidelines can be relied on by the FCC and used directly in enforcement proceedings. It is no longer necessary for FCC staff to duplicate the monitoring and direction finding done by Auxiliary volunteers.

The best course is to try to avoid confrontation. Use extra caution when choosing a frequency to engage your net. Keep your operation legal and strictly above board. If problems persist, apply peer pressure—try to resolve conflicts through negotiation. If you are unsuccessful, contact a representative of the Amateur Auxiliary to the FCC Compliance and Information Bureau. Don't call the FCC—they'll simply refer you to the Auxiliary, per standard operating procedure.

CHAPTER 5

The Bands

R adio amateurs enjoy a wealth of operating space up and down the whole radio spectrum. In fact, they have enjoyed their "own space" since the earliest days of radio, but not before a few hurdles were jumped.

A Naval Battle

In 1909, hundreds of amateur stations outnumbered Navy and commercial stations. Interference and bedlam reigned. The Roberts Bill of 1909, a trust measure sponsored by the US Navy, was one of the first attempts to alleviate the situation. The measure was killed by the Marconi Company, which argued that the interference existed because of commercial operators' antiquated, obsolete equipment not incorporating the most modern tuners. The Marconi Company, along with amateurs, were the only operators using such tuners.

Amateur Radio grew to be a major issue of the day. There were numerous other attempts to snuff out Amateur Radio, some rather devious, and all unsuccessful. In 1912, the government, smarting from its losses, had a new ploy. In new regulations, amateur stations were limited to wavelengths of less than 200 meters. The body of scientific thought at the time held that radio waves became more effective and useful as a direct ratio to their length. The feeling was that amateurs would be dissolved by relegating them to the "useless" spectrum at 200 meters and below. And, the rest is history! Thanks to this "banishment," amateurs were able to become the greatest pioneers of communications science and technology.

Sharing the Bands

Today, there are numerous other radio services vying for pieces of the spectrum pie, all with legitimate purposes. To accommodate them all, the international and domestic regulatory agencies often put them together, taking into consideration priority of importance of each service, and compatibility factors. The result is that, in many instances, we amateurs share our bands with others.

While some of our allocations are *exclusive*, we share with others on the basis of priority. In some cases, we are the *primary* occupant, but share the band with other *primary* occupants on a *coprimary* basis. In this case, we operate on a basis of equality: We must not cause harmful interference to the other *primary* occupants, and they must not interfere with us.

In other cases, Amateur Radio is designated a *secondary* service: We must not cause harmful interference to, and must tolerate interference from, stations in the primary service.

With Whom Do We Share Our Bands?

In the discussions of each of the amateur bands in this chapter, descriptions of the sharing arrangements will be provided. To help you understand the nature of these arrangements, and to help you tell what you're hearing, here is a basic set of

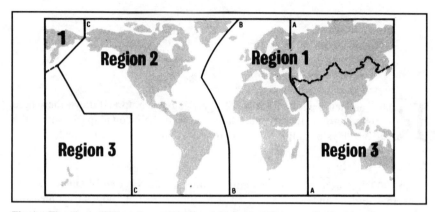

Fig 1—The three ITU regions. Region 1 includes the area limited on the east by line A (lines A, B and C are defined below) and on the west by line B, excluding any of the territory of Iran which lies between these limits. It also includes that part of the territory of Turkey and the former USSR lying outside of these limits, the territory of the Mongolian People's Republic, and the area to the north of the former USSR which lies between lines A and C.
Region 2 includes the area limited on the east by line B and on the west by line C.
Region 3 includes the area limited on the east by line C and on the west by line A, except the territories of the Mongolian People's Republic, Turkey, the former USSR, and the area to the north of the former USSR. It also includes that part of the territory of Iran lying outside of those limits. Lines A, B and C are defined as follows:
Line A extends from the North Pole along meridian 40°E, to parallel 40°N, thence by great-circle arc to the intersection of meridian 60°E and the Tropic of Cancer; thence along the meridian 60°E to the South Pole.
Line B extends from the North Pole along meridian 10°W to its intersection with parallel 72°N; thence by great-circle arc to the intersection of 50°W and parallel 40°N; thence by great circle arc to the intersection of meridian 20°W and parallel 10°S; thence along meridian 20°W to the South Pole.
Line C extends from the North Pole by great-circle arc to the intersection of parallel 65° 30'N with the international boundary in Bering Strait; thence by great-circle arc to the intersection of meridian 165°E and parallel 50°N; thence by great-circle arc to the intersection of meridian 170°W and parallel 10°N, thence along parallel 10°N to its intersection with meridian 120°W; thence along meridian 120°W to the South Pole.
(Thanks RSGB *Operating Manual*)

definitions of some of the other services with which we share our bands:

Fixed Service: A radiocommunication service between specified fixed points.

Fixed-Satellite Service: A radiocommunication service between earth stations at specified fixed points when one or more satellites are used.

Mobile Service: A radiocommunication service between mobile and land stations, or between mobile stations.

Land Mobile Service: A mobile service between base stations and land mobile stations, or between land mobile stations. An example is the communications system for a taxi cab company. The base station at company headquarters dispatches its taxis via radio.

Maritime Mobile Service: A mobile service between coast stations and ship stations, or between ship stations.

Aeronautical Mobile Service: A mobile service between aeronautical stations and aircraft stations, or between aircraft stations.

Broadcasting Service: A radiocommunication service in which the transmissions are intended for direct reception by the general public.

Radionavigation Service: A radiodetermination service for the purpose of navigating a course. Loran systems, for example, guide ships on their chosen courses.

Radiolocation Service: A radiodetermination service for the purpose of locating objects. Radar, for example, is used for tracking movement of ships and planes.

Bands Differ Region To Region

For convenience in organizing frequency allocations for the various services, the International Telecommunication Union has divided the world into three regions: 1, 2 and 3. See Fig 1. North and South America and surrounding waters comprise Region 2.

Frequency allocations for Amateur Radio and other services can differ among ITU Regions. The same sharing rules as outlined above apply. Where, in adjacent regions, a band is allocated to different services of the same category, the basic principle is the equality of right to operate. Stations in a *secondary* service must not cause harmful interference to, and are not protected from interference from, stations in the *primary* service.

Voluntary Band Plans

There's another aspect to sharing the bands: voluntary band plans. Although the FCC rules set aside portions of some bands for specific modes, there's still a need to further organize our space among user groups by "gentlemen's agreements." These agreements usually emerge by consensus of the band occupants, and are sanctioned by a national body like ARRL. They help groups avoid getting in the way of each other. For example, a "DX Window" is set aside on some bands where domestic QSOs are avoided so stations can hear and work weak-signal DX stations. There are others, and we will review them on a band-by-band basis in the following discussion.

A Note about CW

CW is permitted throughout all the bands allocated to amateurs by the

FCC, but check the "gentlemen's agreements" to find out where CW operation is recommended.

Key To Graphs

In the following discussion, graphs are included to illustrate the band restrictions. The key:

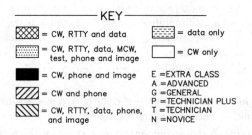

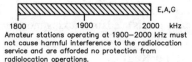

Amateur stations operating at 1900–2000 kHz must not cause harmful interference to the radiolocation service and are afforded no protection from radiolocation operations.

160 Meters: 1800-2000 kHz

Sharing Arrangements: Amateur Radio has *secondary* status to the *primary* radiolocation service (both government and nongovernment) from 1900 to 2000 kHz [2.106; 97.301; 97.303(b),(c)]. Domestically, Amateur Radio is primary from 1800 to 1900 kHz.

A possible dark cloud looms over the horizon with respect to 1900-2000 kHz: A future rule making proceeding in conjunction with expansion of broadcasting into the 1625-1705 kHz band may require some shifting of assignments in this region of the spectrum [US Footnote 290].

Other Regions: In ITU Region 1, radiolocation is the exclusive service at 1800-1810 kHz. Amateur Radio is the exclusive service from 1810-1850 kHz. 1850-2000 kHz is allocated to the fixed and mobile services, with no amateur presence [2.106].

In ITU Region 3, Amateur Radio shares the entire segment 1800-2000 kHz on a *coprimary* basis with fixed, mobile (except aeronautical mobile) and radionavigation services; the radiolocation service is *secondary* [2.106].

License Privileges: General, Advanced and Amateur Extra class amateurs have access to the entire segment, 1800-2000 kHz [97.301(b),(c),(d)].

Mode Privileges: CW, phone, image, RTTY and data modes are permitted across the entire segment [97.305(c)].

Band Planning: The 160-meter band has traditionally been known as the "gentleman's band" or "top band." A voluntary band plan has developed over time to provide coordination of the various operating activities: ARRL recommends use of 1800-1840 kHz for CW, RTTY and other narrow-band modes.

1840-2000 kHz should be reserved for phone, CW, slow-scan television and other wideband modes.

The "DX Window," where only intercontinental QSOs should occur, is 1830-1850. This allows your neighbor to hear DX stations without local interference. Were it not for the voluntary cooperation of all operators on the band in keeping this segment clear, DXing would be difficult and frustrating at best. The band plan for the rest of Region 2, and other regions is outlined in the *ARRL Operating Manual*.

80 METERS

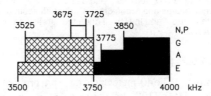

80 Meters: 3500-4000 kHz

Sharing Arrangements: Here in the US, amateurs enjoy *exclusive* operation in the entire segment; the FCC has declined to put any other service here even though it would be permitted to do so under the international regulations. In the rest of Region 2, Amateur Radio enjoys *exclusive* status from 3500-3750 kHz, and shares the rest of the band with foreign fixed and mobile services on a *coprimary* basis [2.106].

Other Regions: In Region 1, Amateur, Fixed and Mobile Service stations share 3500-3800 kHz on a *coprimary* basis. The rest of the band excludes amateur operation. Fixed, land mobile and aeronautical mobile services are the *coprimary* occupants from 3800-3900 kHz. Aeronautical mobile operations have *exclusive* access to 3900-3950 kHz. Broadcasting and fixed stations have the segment 3950-4000 kHz on a *coprimary* basis [2.106].

In Region 3, amateurs share the 3500-3900 kHz segment on a *coprimary* basis with fixed and mobile stations. Amateurs are excluded from 3900-4000 kHz. The aeronautical mobile and broadcast services have access to 3900-3950 kHz on a *coprimary* basis. The broadcasting and fixed services have access to the 3950-4000 kHz segment on a *coprimary* basis.

License privileges: Novice and Technician Plus operators can operate at 3675-3725 kHz, international Morse code only [97.301(e), 97.307(f)(9)]. Generals may operate on 3525-3750, and 3850-4000 kHz [97.301(d)]. Advanced class licensees have access to 3525-3750 kHz, and 3775-4000 kHz [97.301(c)].

Mode privileges: RTTY and data modes are authorized on the non-phone portion only, 3500-3750 kHz. Phone and image emissions are authorized on the voice segment at 3750-4000 kHz [97.305(c)].

All licensees must observe a 200 W PEP power limit when operating in the Novice/Technician Plus subband [97.313(c)].

Band planning: The considerate operator recognizes 3580-3620 kHz as the digital activity area with packet at 3620-3635 kHz [Region 2 band plan, Curacao, 1992]. The RTTY DX channel is 3590 kHz. 3790-3800 kHz is the phone DX window. 3845 kHz is the slow scan television (SSTV) channel. 3885 kHz is the AM calling frequency.

40 METERS

†Phone and image operation is allowed on 7075-7100 kHz by US stations operating in ITU Regions 1 and 3 and by US stations in Puerto Rico, US Virgin Islands and areas of the Caribbean south of 20 degrees north latitude; and in Hawaii and other areas near ITU Region 3, including Alaska

40 Meters: 7000-7300 kHz

Sharing arrangements: The 40-meter band has a big problem: International broadcasting occupies the 7100-7300 kHz band in many parts of the world (Regions 1 and 3). During the daytime when sunspots are high, broadcasting does not cause much interference to US amateurs. However, at night, especially when sunspots are low, the broadcast interference is heavy. Some countries allocate only the 7000-7100 kHz band to amateurs. Others, particularly in our Region 2, allocate 7100-7300 as well, which at times is subject to interference. The result is that there is a great demand for frequencies in the 7000-7100 kHz slot. The effect is that there are two band plans overlaid on each other: ours that spreads out over 7000-7300 kHz and the foreign one that compresses everything into 7000-7100 kHz. Foreign stations typically use the upper part of this segment for phone and the 7035-7045 kHz subband for RTTY. The two basic band plans coexist nicely until propagation permits stations in different regions to hear each other. The consolation is that 7000-7035 kHz is fine hunting ground for domestic and DX CW contacts [2.106].

License privileges: The Novice/Technician Plus subband on 40 meters is 7100-7150 kHz. Novice and Technician Plus operators can use international Morse code only [97.301(e), 97.307(f)(9)].

General class licensees are allowed operation on 7025-7150 kHz, and 7225-7300 kHz [97.301(d)]. Advanced class operators have privileges at 7025-7300 kHz [97.301(c)]. All licensees must limit power output to 200 W PEP on the Novice/Technician Plus subband [97.313(c)].

Mode privileges: RTTY and data modes are allowed on the non-voice frequencies 7000-7150 kHz [97.305(c)].

The phone segment is 7150-7300 kHz [97.305(c)]. Image modes are also permitted here. Phone and Image operation is permitted on 7075 to 7100 kHz for FCC licensed stations in ITU Regions 1 and 3, and by FCC licensed stations in ITU Region 2 west of 130°W longitude or south of 20°N latitude [97.307(f)(11)]. Novice/Technician Plus operators outside of Region 2 may operate CW at 7050-7075 kHz [97.301(e)].

Band planning: Because 40-meter operating patterns in the US are somewhat different than those in the rest of the Region, this band poses a difficult problem. While the US and its possessions in Region 2 are allowed SSB operation only from 7150 kHz and up, the rest of Region 2 uses the band from 7050 (and sometimes even lower) and up for SSB. At the Region 2 meeting in Curacao in 1992, no agreement could be reached that reflected both existing packet activity and the strong desires of other countries for 7050-7100 kHz to be kept free of packet and other digital modes in favor of SSB.

The band segment favored by the rest of the Region (other than the US) for packet, 7040-7050 kHz was not acceptable to the US delegation because of continuing high levels of CW activity. The US delegation made it clear to the conference that a packet segment in the 7040-7050 kHz slot could not be implemented in the US for that reason.

And while Region 2 non-US delegations supported a packet segment at 7100-7120 kHz, this would interfere with US Novice/Technician Plus operators. However, it was deemed that this may not be an insurmountable problem.

The conference set the range 7035-7050 kHz as the digital zone, with packet priority at 7040-7050 kHz for international communication; 7100-7120 kHz packet priority for communications within Region 2. [Region 2 Band Plan, Curacao, 1992.] However, in the US, 7080-7100 kHz is still recognized as the digital area, with 7040 as the RTTY DX channel. Slow-scan TV centers around 7171 kHz. 7290 kHz is recognized as the AM calling frequency.

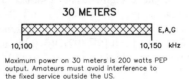

30 METERS

10,100   10,150 kHz

Maximum power on 30 meters is 200 watts PEP output. Amateurs must avoid interference to the fixed service outside the US.

30 Meters: 10,100-10,150 kHz

Sharing arrangements: The Amateur Service is secondary in this band to stations in the Fixed Service outside of the US. You must avoid causing harmful interference to these foreign fixed stations. If you do, you must be prepared to stop transmitting, if necessary, to eliminate the interference [2.106, 97.303(d)].

License privileges: General, Advanced and Amateur Extra class licensees have access to the entire segment, but are limited to 200 W PEP output using CW, RTTY and data emissions only, AMTOR, ASCII and Baudot only, 300 bauds sending limit [97.301(b),(c),(d);97.305(c); 97.307(f)(3); 97.313(c)].

Band Planning: RTTY emissions should be restricted to 10.130-10.140 MHz, with packet operation at 10.140-10.150 MHz.

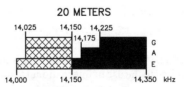

20 METERS

14,025   14,150   14,225
              14,175
14,000   14,150   14,350 kHz

20 Meters: 14.000-14.350 MHz

The 20-meter band is allocated on an *exclusive* basis throughout the world [2.106].

License privileges: No Novice, Technician or Technician Plus privileges are afforded at 20 meters. The General class segments of the band occur at 14.025-14.150 and 14.225-14.350 MHz [97.301(d)]. Advanced class licensees have access to 14.025-14.150 MHz, and 14.175-14.350 MHz [97.301(c)].

Mode privileges: The phone band is 14.150-14.350 MHz [97.305(c)]. Image emissions are permitted on this voice segment also [97.305(c)]. RTTY and data modes are permitted on the non-phone segment 14.000-14.150 MHz.

Band planning: RTTY activity is found between 14.070 and 14.095 MHz with a packet priority segment at 14.095-14.0995 MHz and a packet shared with foreign SSB segment at 14.1005-14.112 MHz. Beacons sponsored by the Northern California DX Foundation can be found at 14.0995-14.1005 MHz. The SSTV channel centers on 14.230 MHz. The AM calling frequency is 14.286 MHz.

17 METERS

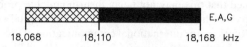

17 Meters: 18.068-18.168 MHz

The 17-meter band was awarded to amateurs on an *exclusive* basis, worldwide, at WARC-79 [2.106]. The entire segment is available to General, Advanced and Amateur Extra class amateurs [97.301(b),(c),(d)]. RTTY and data modes are allowed on 18.068-18.110 MHz [97.305(c)]. RTTY/data modes must be in ASCII, AMTOR or Baudot, 300 bauds limit [97.307(f)(3)]. Phone and image modes are allowed on the rest of the band—18.110-18.168 MHz [97.305(c)].

Band planning: The US band plan calls for CW operation at 18.068-18.100 MHz; the RTTY segment is 18.100-18.105 MHz; and the packet segment is at 18.105-18.110 MHz. Phone, CW and image operation occupies the rest of the band at 18.110-18.168 MHz.

15 METERS

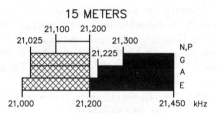

15 Meters: 21.000-21.450 MHz

Another popular DX band, 15 meters is also worldwide amateur *exclusive* [2.106].

License privileges: The Novice/Technician Plus subband is 21.100-21.200 MHz, where they are limited to the use of international Morse code only [97.301(e); 97.307(f)(9)]. General class operators are allowed to use the 21.025-21.200 MHz, and 21.300-21.450 MHz segments [97.301(d)]. Advanced class licensees have 21.025-21.200 MHz and 21.225-21.450 MHz [97.301(c)]. All licensees must limit their power output to no more than 200 W PEP when operating in the Novice/Technician Plus subband [97.313(c)].

Mode privileges: The phone band is 21.200-21.450 MHz [97.305(c)]. Image modes are permitted on this voice segment only [97.305(c)]. RTTY and data modes are permitted on the non-voice segment only: 21.000-21.200 MHz [97.305(c)]. Again, RTTY/data emissions must be in Baudot, AMTOR or ASCII, and sent at not more than 300 bauds [97.307(f)(3)].

Band planning: The Region 2 band plan allows digital operation at 21.070-

21.125 MHz, with packet priority at 21.090-21.125 MHz [Curacao, 1992]. The problem with this arrangement for the US is that it promotes interference to Novice and Technician Plus operators in their subband. So, in the US, the band plan is as follows: RTTY activity is restricted to 21.070-21.090 MHz with packet priority at 21.090-21.100 MHz. The phone band is 21.200-21.450 MHz with an SSTV channel at 21.340 MHz.

12 Meters: 24.890-24.990 MHz

The 12-meter band was made available for use by US amateurs at 0001 UTC on June 22, 1985. It is an amateur exclusive band worldwide [2.106].

License privileges: General, Advanced and Amateur Extra class licensees have access to the entire segment [97.301(b),(c),(d)].

Mode privileges: RTTY and data modes are authorized from 24.890-24.930 MHz, AMTOR, ASCII or Baudot, 300 bauds limit [97.305(c); 97.307(f)(3)]. The phone and image band is 24.930-24.990 MHz [97.305(c)].

Band planning: The band plan calls for CW only at 24.890-24.920 MHz; RTTY/data modes at 24.920-24.925 MHz; packet priority at 24.925-24.930 MHz; and phone, SSTV and CW at 24.930-24.990 MHz [Region 2 band plan, Curacao, 1992].

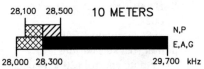

Novices and Technician Plus Licensees are limited to 200 watts PEP output on 10 meters.

10 Meters: 28.0-29.7 MHz

The entire 10-meter band is amateur *exclusive* worldwide [2.106]. The Novice/Technician Plus subband is 28.1-28.5 MHz [97.301(e); 97.307(f)(10)]. Novice and Technician Plus licensees must limit their power to no more than 200 W PEP output [97.313(c)(2)]. All other licensees have access to the entire band [97.301(b)(c)(d)].

RTTY and data modes are permitted at 28.0-28.3 MHz [97.305(c)]. RTTY/data emissions must be limited to AMTOR, Baudot or ASCII code; the sending speed limit is 1200 bauds [97.307(f)(4)].

The phone band is 28.3-29.7 MHz [97.305(c)]. Image emissions are allowed here, too. FM voice is effectively limited to frequencies above 29.0 MHz [97.307(f)(1)] and repeaters are permitted from 29.5-29.7 MHz [97.205(b)].

Band planning: The Region 2 band plan recognizes RTTY/data operation at 28.070-28.189 MHz, with packet priority at 28.120-28.189 MHz [Curacao, 1992]. SSTV activity centers around 28.680 MHz. Beacons are found between 28.190 and 28.225 MHz. The phone band is 28.300-29.300 MHz. AM operation is found at 29.000-29.200 MHz.

Satellite downlinks exist between 29.30 and 29.510 MHz. Repeater inputs are found from 29.520 to 29.580 MHz and outputs from 29.620-29.680 MHz. The FM simplex calling frequency is 29.600 MHz.

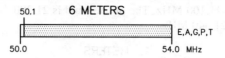

50.1   **6 METERS**

E,A,G,P,T

50.0   54.0   MHz

6 Meters: 50-54 MHz

Six meters is a popular VHF band. In Regions 2 and 3, it is amateur *exclusive*. It is exclusively allocated for broadcasting in Region 1 [2.106]. The entire band is available for use by all licensees except Novices [97.301(a)].

There is a CW exclusive subband from 50.0 to 50.1 MHz. MCW, phone, image, RTTY and data modes are permitted throughout the rest of the band, 50.1-54.0 MHz [97.305(c)].

RTTY and Data modes used in sending specified codes may be transmitted at speeds up to 19.6 kilobauds [97.307(f)(5)]. Experimental (unspecified) codes are allowed—the bandwidth must not exceed 20 kHz [97.307(f)(5)].

Band planning: Activities in the 6-meter band are depicted in the following band plan [ARRL, July 1991].

50.0-50.1	CW, beacons
50.060-50.080	beacon subband
50.1-50.3	SSB,CW
50.10-50.125	DX window
50.125	SSB calling
50.3-50.6	all modes
50.6-50.8	nonvoice communications
50.62	digital (packet) calling
50.8-51.0	radio remote control (20-kHz channels)
51.0-51.1	Pacific DX window
51.12-51.48	repeater inputs (19 channels)
51.12-51.18	digital repeater inputs
51.5-51.6	simplex (6 channels)
51.62-51.98	repeater outputs (19 channels)
51.62-51.68	digital repeater outputs
52.0-52.48	repeater inputs (except as noted; 23 channels)
52.02, 52.04	FM simplex
52.2	TEST PAIR (input)
52.5-52.98	repeater outputs (except as noted; 23 channels)
52.525	primary FM simplex
52.54	secondary FM simplex
52.7	TEST PAIR (output)
53.0-53.48	repeater inputs (except as noted; 19 channels)
53.0	remote base FM simplex
53.02	simplex
53.1, 53.2,	radio remote control
53.3, 53.4	
53.5-53.98	repeater outputs (except as noted; 19 channels)

53.5, 53.6	radio remote control
53.7, 53.8	
53.52-53.9	simplex

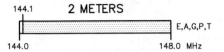

2 Meters: 144-148 MHz

The most popular amateur VHF band is 2 meters, where much FM repeater activity occurs. The band is amateur *exclusive* in Region 2.

In Region 1, amateurs are limited to 144-146 MHz, but it is an *exclusive* allocation. Fixed and mobile stations occupy 146-148 MHz in this region.

In Region 3, amateurs enjoy 144-146 exclusively, and share 146-148 MHz with fixed and mobile stations on a *coprimary* basis [2.106].

There is a CW-only segment from 144.00 to 144.10 MHz. All of the remaining emissions are permitted above 144.10 MHz [97.305(c)].

All licensees except Novices are permitted to use the entire segment [97.301(a)].

The following band plan explains typical operation on the 2-meter band:

144.00-144.05	EME (CW)
144.05-144.10	General CW and weak signals
144.10-144.20	EME and weak-signal SSB
144.200	National calling frequency
144.20-144.275	General SSB operation
144.275-144.300	Propagation beacons
144.30-144.50	New OSCAR subband
144.50-144.60	Linear translator inputs
144.60-144.90	FM repeater inputs
144.90-145.10	Weak signal and FM simplex
	(145.01,03,05,07,09 are widely used for packet radio)
145.10-145.20	Linear translator outputs
145.20-145.50	FM repeater outputs
145.50-145.80	Miscellaneous and experimental modes
145.80-146.00	OSCAR subband
146.01-146.37	Repeater inputs
146.40-146.58	Simplex
146.61-147.39	Repeater outputs
147.42-147.57	Simplex
147.60-147.99	Repeater inputs

1.25-Meters: 222-225 MHz

Sharing arrangements: In the US, amateurs have *exclusive* use of the 222-225 MHz band. There are no amateur allocations outside Region 2 [2.106].

All licensees have access to the entire band 222-225 MHz [97.301(a)(f)]. All modes are permitted [97.305(c)]. A weak signal "guard band" is found at 222.0-222.15 where repeater and auxiliary operation is prohibited [97.201(b); 97.205(b)]. Novice licensees must limit their power to no more than 25 W PEP output [97.313(d)].

1.25 METERS

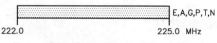

222.0   225.0 MHz

         E,A,G,P,T,N

Novices are limited to 25 watts PEP output
from 222 to 225 MHz.

Band planning [ARRL, July 1991].

222.0-222.15	Weak-signal modes
222.0-222.025	EME
222.05-222.06	Propagation beacons
222.1	SSB & CW calling frequency
222.10-222.15	Weak-signal CW & SSB
222.15-222.25	Local coordinator's option; weak signal, ACSB, repeater inputs, control
222.25-223.38	FM repeater inputs only
223.40-223.52	FM simplex
223.5	Simplex calling frequency
223.52-223.64	Digital, packet
223.64-223.7	Links, control
223.71-223.85	Local coordinator's option, FM simplex, packet, repeater outputs
223.85-224.98	Repeater outputs only

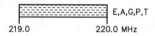

219.0   220.0 MHz

     E,A,G,P,T

219-220 MHz: The FCC has allocated 219-220 MHz to amateur use on a secondary basis. This allocation is for fixed digital message forwarding systems operated by all licensees except Novices. Amateur operations must not cause interference to, and must accept interference from, primary services in this and adjacent bands. Amateur stations are limited to 50 W PEP output and 100 kHz bandwidth. Automated Maritime Telecommunications Systems (AMTS) stations are the primary occupants in this band. Amateur stations within 398 miles of an AMTS station must notify the station in writing at least 30 days prior to beginning operations. Amateur stations within 50 miles of an AMTS station must get permission in writing from the AMTS station before beginning operations. ARRL Headquarters maintains a database of AMTS stations. The FCC requires that amateur operators provide written notification including the station's geographic location to the ARRL for inclusion in a database at least 30 days before beginning operations.

70 CENTIMETERS

420.0   450.0 MHz

     E,A,G,P,T

70-cm Band: 420-450 MHz

In the US, amateurs share the band with Government radiolocation (Radar). The Amateur Radio Service is the secondary service, and therefore must not inter-

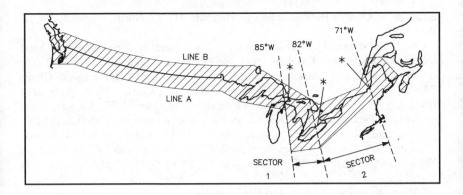

fere with the priority Government stations. We must also tolerate interference from and can't cause interference to Government radiolocation stations [97.303(b)].

The 420-430 MHz segment is allocated to the fixed and mobile (except aeronautical mobile) services in the international allocations table on a primary basis, worldwide. We are not to cause harmful interference to these stations, nor are we protected from interference from these stations [97.303(f)(2)].

Amateur operation in the 420-430 MHz portion of the band is not permitted north of Line A (see figure) [97.303(f)(1)].

The FCC has allocated portions of the band 421-430 MHz to the Land Mobile Service within a 50-mile radius centered on Buffalo, Detroit and Cleveland. Amateur stations south of Line A in the vicinities of these cities may continue to operate in the 421-430 MHz spectrum as long as they do not cause interference to and accept interference from land mobile or government radiolocation users [US Footnote 230, 2.106].

Additionally, 50 W PEP output power limitations apply to amateurs operating within circles centered on designated military installations in the US. Exceptions may be granted when expressly authorized by the FCC after mutual agreement, on a case-by-case basis, with the FCC Engineer-in-Charge and the area's military area frequency coordinator. An earth or telecommand station may, however, transmit on the 435-438 MHz segment with a maximum of 611 W ERP without the authorization required. The transmitting antenna must be pointed above the horizon by at least 10° [97.313(f); US Footnote 7, 2.106].

The affected areas are:

1) Portions of New Mexico and Texas bounded on the south by latitude 31° 45' North, on the east by longitude 104°00' West, on the north by latitude 34°30' North, and on the west by longitude 107°30' West.

2) The entire state of Florida including the Key West area and the areas enclosed within a 200-mile radius of Patrick Air Force Base, Florida (latitude 28°21' North, longitude 80°43' West), and within a 200-mile radius of Eglin AFB, Florida (latitude 30°30' North, longitude 86°30' West);

3) The entire state of Arizona;

4) Those portions of California and Nevada south of latitude 37°10' North, and the areas enclosed within a 200-mile radius of the Pacific Missile Test Center, Point Mugu, California (latitude 34°09' North, longitude 119°11' West).

5) In the state of Massachusetts within a 160-kilometer (100-mile) radius around

locations at Otis AFB, Massachusetts (latitude 41° 45' North, longitude 72°32' West).

6) In the state of California, within a 240-kilometer (150-mile) radius around locations at Beale AFB, California (latitude 39°08' North, longitude 121° 26' West).

7) In the state of Alaska, within a 160-kilometer (100-mile) radius of Clear, Alaska (latitude 64°17' North, longitude 149°10' West), (the military area frequency coordinator for this area is located at Elmendorf AFB, Alaska).

8) In the state of North Dakota, within a 100-kilometer (100-mile) radius of Concrete, North Dakota (latitude 48°43' North, longitude 97°54' West). (The military area frequency coordinator for this area can be contacted at: HQSAC/SXOE, Offutt AFB, Nebraska, 68113.) [97.313(f); US Footnote 7, 2.106.]

When operating at 449.5-450 MHz, amateurs must not cause interference to, and must accept interference from space operation service and space research service stations or government or non-government space telecommand stations [US Footnote 87, 2.106].

License privileges: All amateurs except Novices have access to the entire segment 420-450 MHz [97.301(a)].

Mode privileges: All mode privileges are authorized across the entire band [97.305(c)]. The RTTY and data mode speed limit is 56 kilobauds [97.307(f)(6)]. Experimental codes are allowed on this segment; the maximum bandwidth allowed is 100 kHz [97.307(f)(6)].

Band planning:

420.00-426.00	ATV repeater or simplex with 421.25 MHz video carrier, control links, and experimental
426.00-432.00	ATV simplex with 427.250-MHz video carrier frequency
432.00-432.070	EME (Earth-Moon-Earth)
432.07-432.10	Weak-signal CW
432.100	70-cm calling frequency
432.10-432.30	Mixed-mode and weak-signal work
432.30-432.40	Propagation beacons
432.40-433.00	Mixed-mode and CW work
433.00-435.00	Auxiliary/repeater links
435.00-438.00	Satellite only (internationally)
438.00-444.00	ATV repeater input with 439.250-MHz video carrier frequency and repeater links
442.00-445.00	Repeater inputs and outputs (local option)
445.00-447.00	Shared by auxiliary and control links, repeaters and simplex (local option); (446.0: national simplex frequency)
447.00-450.00	Repeater inputs and outputs

33 CENTIMETERS

E,A,G,P,T

902.0   928.0 MHz

33-cm Band: 902-928 MHz

Sharing arrangements: In the US, amateurs are *secondary* to the government radiolocation (Radar) stations and we can't cause interference to and must accept interference from those stations [2.106; 97.303(b)].

Amateurs may encounter emissions from "Industrial, Scientific and Medical" (ISM) equipment. Examples are dialysis machines and manufacturing equipment. We must not cause interference to, and must accept interference from, these applications [97.303(g)(1)].

We must also accept interference from microwave ovens operating at 915 MHz, manufactured before January 1, 1980 [US Footnote 215, 2.106], and from the Location and Monitoring Service (LMS). We must not cause interference to LMS systems, which operate across the entire band, 902-928 MHz [US Footnotes 218 and 275; 2.106; 97.303(g)(1)].

Amateurs located in the states of Colorado and Wyoming, bounded on the south by latitude 39° North, on the north by latitude 42° North, on the east by longitude 105° West, and on the west by longitude 108° West may not transmit on this band except in the segments 902.0-902.4, 902.6-904.3, 904.7-925.3, 925.7-927.3, and 927.7-928.0 MHz only [97.303(g)(1); July 2, 1990, waiver].

Amateurs located in the states of Texas and New Mexico bounded on the south by 31° 41' North, on the north by latitude 34°30' North, on the east by longitude 104°11' West, and on the west by longitude 107°30' W may not transmit on this band [US Footnote 267, 2.106; 97.303(g)(2)]. In addition, amateurs outside these boundaries but within 150 miles of the White Sands Missile Range must limit their peak envelope power to 50 watts [97.313(g)].

License privileges: All licensees except Novices have full access to the segment [97.301(a)].

Mode privileges: All modes may be used on the entire band [97.305(c)].

Band planning [ARRL, July 1989]:

902.0-903.0	Weak signal [902.1 calling frequency]
903.0-906.0	Digital [903.1 alternate calling frequency]
906.0-909.0	FM repeater outputs
909.0-915.0	ATV
915.0-918.0	Digital
918.0-921.0	FM repeater inputs
921.0-927.0	ATV
927.0-928.0	FM simplex and links

23 CENTIMETERS

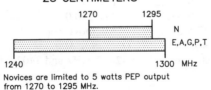

Novices are limited to 5 watts PEP output from 1270 to 1295 MHz.

23-cm Band: 1240-1300 MHz

Again, all emission privileges are allowed on this band, and are available to all licensees except Novices [97.305(c); 97.301(a),(f)]. Novice operators may use the 1270-1295 MHz subband and are limited to 5 watts PEP [97.301(f); 97.313(e)]. This band, as are virtually all amateur bands between 222 MHz and 24.250 GHz (24.00-24.05 GHz is an exception), is allocated to amateurs on a secondary, non-interference basis, to the primary Government Radiolocation services [2.106; 97.303(b)].

Band planning [ARRL, January 1985]:

1240-1246	ATV #1
1246-1248	Narrow-bandwidth FM point-to-point links and digital, duplexed with 1258-1260.
1248-1252	Digital Communications
1252-1258	ATV #2
1258-1260	Narrow-bandwidth FM point-to-point links and digital, duplexed with 1246-1252
1260-1270	Satellite uplinks, reference WARC '79
1260-1270	Wide-bandwidth experimental, simplex ATV
1270-1276	Repeater inputs, FM and linear, paired with 1282-1288. (239 pairs, every 25 kHz, eg, 1270.025, 050, etc)
1271/1283	Non-coordinated test pair
1276-1282	ATV #3
1282-1288	Repeater outputs, paired with 1270-1276
1288-1294	Wide-bandwidth experimental, simplex ATV
1294-1295	Narrow-bandwidth FM simplex services, 25-kHz channels
1294.5	National FM simplex calling frequency
1295-1297	Narrow bandwidth weak-signal communications (no FM)
1295.0-1295.8	SSTV, FAX, ACSSB experimental
1295.8-1296.0	Reserved for EME, CW expansion
1296.0-1296.05	EME-exclusive
1296.07-1296.08	CW beacons
1296.1	CW, SSB calling frequency
1296.4-1296.6	Crossband linear translator input
1296.6-1296.8	Crossband linear translator output
1296.8-1297.0	Experimental beacons (exclusive)
1297-1300	Digital communications

Above and Beyond

All modes and licensees (except Novices) are authorized on the following bands [97.301(a)]:

*2300-2310 MHz	10.0-10.5 GHz	119.98-120.02 GHz
*2390-2450	24.0-24.25	142-149
3300-3500	47.0-47.2	241-250.0
5650-5925	75.5-81.0	All above 300

For band plans, see *The ARRL Operating Manual*.

*Amateurs are primary on the 2390-2400 MHz and 2402-2417 MHz subbands.

CHAPTER 6

Repeaters, Auxiliary Stations and Station Control

M ost sections of Part 97 require interpretation, but none have as many different interpretations as the rules governing repeaters and auxiliary stations. The most important thing here is that the spirit of the rules is as important as the letter. As you read the rules, keep in mind that they are there to give guidance and promote operating practices that do not cause harmful interference. They are not meant to prohibit specific activities. Ask yourself if your on-the-air operations promote the spirit of cooperation that is so important in a self-regulating radio service. Remember the first guideline in The Amateur's Code: "The radio amateur is CONSIDERATE...never knowingly operates in such a way as to lessen the pleasure of others." If you follow this all-important principle, you will almost always find that your activities are permitted in Part 97.

Control

Before dealing with the specialized operating modes, we must discuss the concept of control. The terms *control operator* and *control point* were introduced in Chapter 3, but the FCC specifically states that control of an Amateur Radio station can be done in one of three very different ways. Every station, regardless of the type of control used, must have at least one control point [97.109(a)]. Any station may be locally or remotely controlled [97.109(b), (c)].

Local Control

This is the simplest kind of control. The FCC defines this as "the use of a control operator who directly manipulates the operating adjustments in the station to achieve compliance with the FCC Rules" [97.3(a)(28)]. When you sit in front of your rig and twiddle the knobs, that's local control.

Remote Control

Remote control, logically enough, involves control from a point that is remote from the station being controlled. The FCC calls it "the use of a control operator who indirectly manipulates the operating adjustments in the station through a control link to achieve compliance with the FCC Rules" [97.3(a)(36)]. The key words here are *indirectly* and *control link*. If you wire a control box so you can control your station from another room, that's remote control. If you build a DTMF decoder so you can call your shack from a pay phone and key your transmitter, that's remote control, too. The control link is whatever you use to bring the control commands from your hand to the radio. It can be wire, phone line, radio or whatever. If you control your station *indirectly*, it's *remote control*.

Another FCC definition applies in remote control situations: *Telecommand* is "a one-way transmission to initiate, modify or terminate functions of a device at a distance" [97.3(a)(41)].

The rules contain specifications on remote control and telecommand operation:

- Provision must be incorporated to limit transmissions to no more than three minutes if the control link fails. This is a common-sense provision. If the control link fails while your transmitter is keyed, the transmitter could be seriously damaged (not to mention the interference you'd cause) if there was no three-minute timer [97.213(b)].
- The station must be protected so that unauthorized transmissions cannot be made, whether willfully or negligently. This refers to providing safeguards on your remotely controlled station so it cannot be used by unauthorized operators. Most remote station operators use DTMF tones or CTCSS systems to limit access only to those people who know the codes. You are responsible for the transmissions from your remote station, just as you are responsible for your home station [97.213(c)].
- A photocopy of the station license and a label with the name, address and telephone number of the station licensee and at least one designated control operator must be posted at the station site. This enables the FCC to find a responsible person, if necessary. If someone uses your remote station to make unauthorized transmissions, the FCC may have to use direction-finding equipment to locate the transmitter. When they travel to the station site, they want to find a card listing your name so they can come knock on your door [97.213(d)].
- Control links may be wire (a telephone line, for example) or radio. The FCC says that if a radio control link is used, the station where the control commands are entered for relay to the remotely controlled station is an *auxiliary station* [97.213(a)]. The FCC defines an auxiliary station as "an amateur station transmitting communications point-to-point within a system of co-operating amateur stations [97.3(a)(7)].

Automatic Control

The last type of control is automatic control: "The use of devices and procedures for control of a station when it is transmitting so that compliance with the FCC Rules is achieved without the control operator being present at a control point" [97.3(a)(6)]. This is "hands-off" operation; there's nobody home (at the control point). If the repeater is always active and retransmits anything that comes along, that's automatic control. Most repeaters operate at least part of the time under automatic control.

Not all amateur stations can be operated by automatic control. Only space stations, repeaters, beacons (in certain segments), and auxiliary stations can be operated automatically [97.109(d)]. In addition, stations transmitting RTTY/data emissions above 50 MHz or in the following segments may be operated automatically: 3.620-3.635 MHz; 7.100-7.105 MHz; 10.140-10.150 MHz; 14.0950-14.0995 MHz; 14.1005-14.1120 MHz; 18.105-18.110 MHz; 21.090-21.100 MHz; 24.925-24.930 MHz; and 28.120-28.189 MHz [97.221 (b)]. See Chapter 13 for more details on HF automatic operation; space stations and beacons are discussed in Chapter 8.

Repeaters

The FCC Rules define a repeater as "an amateur station that simultaneously retransmits the transmission of another amateur station on a different channel or channels" [97.3(a)(37)]. Only repeaters, auxiliary and space stations may automatically retransmit the radio signals of other amateur stations [97.113(f)].

A repeater normally consists of a transmitter, a receiver, an antenna, a control box and a duplexer (this allows the use of a single antenna for transmitting and receiving). Operation generally occurs on the VHF amateur bands, although there is 10-meter repeater activity and repeaters are becoming more common on the UHF and microwave bands. The most popular band for repeaters is 144 MHz (2 meters), with 222 MHz and 450 MHz following in popularity. FM is widely used.

A typical machine sits on top of a mountain or tall building and retransmits signals from small hand-held or mobile rigs. The result is an increase in communications coverage for the user—some repeaters extend coverage to entire states and more.

A repeater may be part of a cooperating system of repeaters, linked by wireline or auxiliary stations, extending the geographic range of the system far beyond that of a single repeater. Except for the geographic range involved, the same rules apply to the system as to individual repeaters.

Repeater Control

Because repeaters are generally found on hilltops and tall buildings, local control is not practical. Most repeaters operate under automatic control and have a control link. This link allows the repeater to be disabled by remote control if necessary and the control operator to enable or disable various repeater functions. Although the control operator is not required to be present at a control point under automatic control, it is still the station licensee's responsibility to see that the repeater operates properly at all times [97.103(a)]. The repeater owner should prevent unauthorized tampering by using procedures and devices such as unpublished control-link frequencies and padlocks on the repeater housing [97.3(a)(6)]. Make sure you get the word quickly if something is wrong and have quick access to the repeater shut-down function. Protect your feed lines and antennas whenever possible. The FCC states that "limiting the use of a repeater to only certain user stations is permissible" [97.205(e)]. This allows you to implement subaudible tone security systems, if required to protect the security of your repeater.

Control During Autopatch Operation

A third party participates in amateur communications when you make a phone call with a repeater autopatch. This means that a voice repeater must be locally or remotely controlled and a control operator must be present at the local or remote control point when the repeater autopatch is active [97.115(b)(1); 97.109(e)]. There is more discussion of autopatch and third-party questions in Chapter 7.

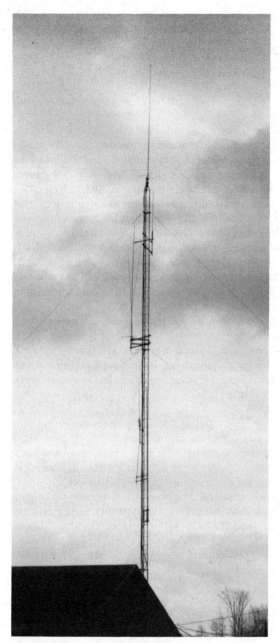

The W1KKF repeater, located at the firehouse atop Cook Hill in Wallingford, Connecticut, allows mobile operators to communicate throughout central Connecticut. It provides communications during emergencies for both local and area Emergency Operations Centers. *(photo by WA1CCQ)*

"Ancillary Functions"

The rules state that "ancillary functions of a repeater that are available to users on the input channel are not considered remotely controlled functions of the station" [97.205(e)].

The FCC is making a distinction between functions made available by the control operator to regular users (a crossband link, time-of-day report or autopatch, for example) and functions reserved exclusively for control operators in effecting basic control of the repeater. These basic functions (turning the repeater on and off, for example) should be performed on frequencies where auxiliary operation is permitted, away from the input frequency.

A more subtle distinction must also be made: Although it is proper for John Q. Ham, an average user, to turn on the autopatch by sending the proper touchtones—using the patch to make a call, then turning it back off, all on the input frequency—the act of changing the command state of the repeater to enable the autopatch function, thus making it available to users in the first place, is a remote-control function reserved for the control operator.

The same distinction applies to crossband links and other functions. If the control operator enables (makes available) a crossband link for use by the average user by changing the repeater's command state through the control link, then users, including Novices on a 222-MHz repeater, may turn on and use the crossband link on the repeater input frequency to communicate with other hams on 2 meters. Changing the command state to allow users access to the crossband link is a *control function*. The average user accessing the crossband link on the repeater input frequency is an *ancillary function*.

The point of this discussion of repeater control is that you are responsible for the emissions of your repeater at all times [97.103(a)]. You must be able to shut off the repeater if there is a problem. If your only means of control is on the repeater input frequency, there may be times when jamming or unintentional interference make this impossible. You should provide an alternate means of control, by telephone or via a radio link on another frequency.

Repeater Station Identification

Repeaters may be identified by CW or voice [97.119(b)]. If an automatic device is used for the CW ID, the sending speed cannot exceed 20 words per minute [97.119(b)(1)]. The FCC no longer requires that repeaters sign "/R" on CW or "repeater" on voice. Repeaters can also be identified using ASCII, Baudot or AMTOR when that particular code is used for all or part of the communication or when the communication is transmitted in any digital code above 50 MHz [97.119(b)(3)]. This is why packet TNCs do not need to identify with CW. Amateur fast-scan television repeaters may identify with video as long as the US standard 525-line system is used [97.119(b)(4)].

Who May Use a Repeater

Repeater owners may limit the use of their repeaters. The FCC recognizes that a repeater is an Amateur Radio station, licensed to its owner, and that repeater owners may limit the use of their stations [97.205(e)]. This specifically permits operation of a "closed" repeater using CTCSS or other systems to limit access. A repeater does not have to be listed as "closed" to limit access, however.

Another important point is that a repeater user is not usually the control operator. Thus, a Novice licensee may use a 222-MHz to 144-MHz linked repeater system. As long as the input of the repeater is on a frequency authorized for Novice use, Novices may use the system, even though their signals are being retransmitted on a band where they are not authorized to operate directly. In this case, the Novices are control operators only of their stations on 222 MHz. They are not transmitting on 2 meters; the repeater station is and it is operating under the trustee's or control operator's license privileges. The same situation applies to Technician and Technician Plus licensees operating above 50 MHz using a repeater with its output on 10 meters.

Other Repeater Rules

- A Novice operator may not be the licensee or control operator of a repeater [97.205(a)]. Any other licensed operator may put up a repeater, as long as the outputs are all on frequencies authorized to the trustee.
- Specific frequencies are available for repeater operation. Repeaters may operate on any frequency authorized to the Amateur Radio service above 29.5 MHz, except for 50.0-51.0, 144.0-144.5, 145.5-146.0, 222.0-222.15, 431.0-433.0 and 435-438 MHz [97.205(b)].
- Two repeater owners must work together to solve an interference problem

between them, unless one repeater is coordinated and the other is not. The licensee of an uncoordinated repeater has primary responsibility for solving an interference problem. This is the FCC's way of recognizing the importance of frequency coordination [97.205(c)].

• The control operator of a repeater that inadvertently retransmits communications that violate the rules in Part 97 is not held accountable for the violative communications [97.205(g)].

Simplex Repeaters

The availability of digital devices which can store a limited amount of audio has led to the development of devices called *simplex repeaters*. These devices are used to extend the range of low power rigs, such as H-Ts, by storing the signal transmitted on a channel, and then retransmitting it through a more powerful and/ or better situated transmitter.

According to the rules, however, such a device is not a repeater. It does not *simultaneously* retransmit the signals of another station, nor does it retransmit the signals on a *different* channel or channels, as specified in the FCC's definition of a repeater. Since it isn't a repeater, such a device cannot be operated under automatic control. If a control operator is present and controlling the device, either locally or by remote control, then it can be used. It cannot be left unattended, however.

Digipeaters

With the introduction of packet radio technology in the amateur service, another device has seen widespread use—the digipeater. A digipeater receives, processes and retransmits processed packets of data. Digipeaters relay digital transmissions from other stations and extend the available range for packet communications.

There is a distinction between ordinary repeaters and digipeaters. Digipeaters confirm, process and then pass along data packets; *they do not automatically retransmit signals without changing the information contained, as repeaters do*. As part of the process, digipeaters change data bits, so information transmitted is different from that received. Digipeaters are not repeaters, so repeater rules do not apply to their operation. For example, Novices are permitted to be control operators for digipeaters, but not for voice repeaters. Digipeaters are not limited to the repeater subbands. No station may be automatically controlled while transmitting third party communications, except a station participating as a forwarding station in a message forwarding system [97.109(e)]. A message forwarding system is a group of amateur stations participating in a voluntary, cooperative, interactive arrangement where communications are sent from the control operator of an originating station to the control operator of one or more destination stations by one or more forwarding stations [97.3(a)(29)]. The originating station and the first forwarding station must accept responsibility for any messages transmitted [97.219].

Digipeaters are remotely or automatically controlled stations that receive, process and send data emissions; they are not actually repeaters.

Auxiliary Stations

When an amateur station is controlled remotely by radio, there is another station involved—the station doing the controlling. This station is called an auxiliary station. Here's how the FCC defines an auxiliary station: "An amateur station transmitting communications point-to-point within a system of cooperating amateur stations" [97.3(a)(7)]. There are a few important rules that apply to auxiliary stations:

- Novice licensees may not put auxiliary stations on the air or be control operators for existing auxiliary stations. Amateurs with any other class of license may put auxiliary stations on the air or be control operators for auxiliary stations [97.201(a)].
- Auxiliary stations may only transmit in ham bands above and including 222 MHz, except in the segments 222.0-222.15, 431-433 and 435-438 MHz [97.201(b)].
- Licensees of auxiliary stations causing interference to each other are equally responsible for solving the interference, except where one station is coordinated and the other is not [97.201(c)]. Again, this is the FCC's way of recognizing the importance of frequency coordination. We'll discuss this critical topic later in the chapter.
- An auxiliary station may be automatically controlled and may send one-way transmissions [97.201(d),(e)].

The easiest way to explain the operation of an auxiliary station is to use examples.

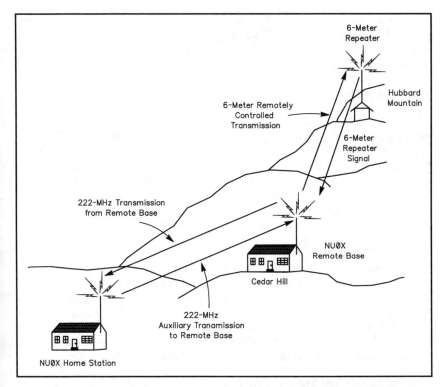

Fig 1—NUØX, Jay, uses an auxiliary station to remotely control the NUØX remote base atop Cedar Hill. This enables him to check into the nightly Rock-climbing/SCUBA/Kayaking net on the Hubbard Mt. 6-meter repeater, where he discusses karate with Rick, K1CE, and Norm, WA1CCQ.

Remote Bases

Auxiliary stations are used in what have come to be known as "remote bases." A remote base is a station set up, usually in a good location, to relay signals from other stations. One example might be a 50-MHz transceiver linked to a 222-MHz transceiver. Users on 222 MHz have their signals retransmitted on 50 MHz. 50-MHz signals are received and retransmitted on 222 MHz. This is shown in Fig 1. When you are transmitting control commands and audio on 222 MHz, your station is an auxiliary station. The 50-MHz transmitter is not an auxiliary station; it's an ordinary station under remote control. The 222-MHz transmitter at the remote site that relays 50-MHz signals to you is an auxiliary station. It's relaying signals from another band.

The remotely controlled transmitter or remote base is not a repeater, as defined by the FCC Rules. The operator of a remote base is the control operator of that station and must not operate the remotely controlled transmitter beyond the privileges allowed by his or her license class. For example, while the 2-meter signals of a Technician class licensee may be repeated on 10 meters if the license class of the repeater's control operator permits operation on the 10-meter repeater subband, a Technician may not operate a remote base with an output on a "non-Technician" frequency.

"Crossband Repeaters"

Modern dual band VHF/UHF rigs often have the capability to do crossband linking. When operating in this mode, they are called "crossband repeaters," but in fact are usually operated as remote bases. Generally, they are used to allow an operator to access a repeater with a hand-held radio (H-T) from a location where he or she would normally not be able to do so. For example, a hiker in a remote location might leave his car where his dual band mobile rig can access a 2-meter repeater. Leaving the mobile rig on, he then takes his UHF H-T with him, and can access the 2-meter repeater via his mobile rig. This is okay as long as two conditions are met: (1) The user links to his crossband rig on the UHF side. This serves as his control link, so is an auxiliary station and must be above 222.15 MHz. (2) The unattended mobile rig is capable of identifying on its UHF output. Since this is a remote base, the user's ID over the UHF link to the dual band radio serves also to ID the VHF side; in the other direction, however, there is no way for the user to identify the UHF output of the remote base, so some form of automatic ID must be employed. Few manufacturers include this ID capability in their rigs.

Simplex Autopatch

The simplex autopatch ("simpatch") is a device connected to a home station that can connect you to a telephone line. A simplex autopatch is controlled by an auxiliary station, so these devices must be used on frequencies authorized for auxiliary stations (above 222 MHz, except 222.0-222.15, 431-433 MHz and 435-438 MHz). You transmit control commands to the simpatch; this makes your station an auxiliary station. The simpatch transmits audio from the phone line; it's also an auxiliary station. This is treated in more detail in Chapter 7.

Relaying Signals in a Repeater System

Auxiliary stations are used to relay signals between separate parts of a repeater system. Having the repeater receiver and transmitter at one site can cause problems. These can sometimes be solved with filters and duplexers, but an alternative

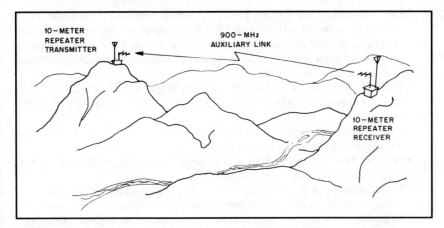

Fig 2—Another use for an auxiliary station. Here, a 900-MHz link relays signals from a 10-meter repeater receiver to the repeater transmitter, located a few miles away.

is to separate the transmitter and receiver. Several 10-meter repeater systems use this technique. An auxiliary link on another band can then be used to relay signals from the receiver to the transmitter. This is shown in Fig 2. In addition, auxiliary stations are used to link separate repeaters into a network spanning a larger geographic area than can be covered by one repeater.

Frequency Coordination

Coordinating one's choice of frequencies for a repeater has never been a licensing requirement. By early 1985, however, demand for frequencies had grown to the point where even an FCC predisposed to "deregulation" decided something needed to be done. In February 1985, the FCC (in PR Docket 85-22) stated:

> Most of the reported cases of amateur repeater-to-repeater interference appear to involve one or more noncoordinated repeaters. In the past two years we have had to resolve amateur repeater-to-repeater interference disputes in which at least one noncoordinated repeater was involved and in which the parties to the dispute could reach no amicable solution. We attribute the growing number of instances of amateur repeater-to-repeater interference and the need for increased intervention in these matters to the mounting pressure which develops as the desirable repeater frequencies become fully assigned.
>
> When we have intervened in such interference disputes, we have…favored the repeater operating in accordance with the recommendation of a local frequency coordinator.
>
> Comments are sought with regard to whether the FCC should recognize a single national frequency coordinator for the Amateur Radio service. Such a coordinator could be an individual national organization or an umbrella organization comprised of local coordination groups. We would expect this type of organization to promulgate coordination (and de-coordination) standards, to facilitate the use of advancing technology, to consider alternative frequency assignments, to consider frequency spacing and repeater separation distances and, in the case of an umbrella organization, to advise local coordinators.

In its comments to PR Docket 85-22, the ARRL proposed these guidelines:

- Preferred status in instances of harmful repeater-to-repeater interference should be granted to amateur repeater operators who have implemented the recommendation of their local or regional volunteer frequency coordinator and are thereby coordinated.
- Frequency coordination should be strongly urged for all amateur stations in repeater or auxiliary operation in any geographical area served by a frequency coordinator.
- The commission should not consider alternatives to frequency coordination nor mandate methods of coordination.
- The use of modern technological innovation…should be encouraged, but not substituted for frequency coordination.
- The commission should not recognize a single entity, such as a national frequency coordinator, for amateur repeater operation. Such coordination activities should be performed by local or regional volunteer frequency coordinators, with appropriate support to these coordinators provided by the League.

In its Report and Order to PR Docket 85-22, the commission stated:

Repeater operation in the amateur service inherently requires operation on established frequencies. Amateur repeater operation is not frequency agile, as are other types of amateur station operation. As a result, most amateur operators have been willing to voluntarily cooperate to avoid interference to frequencies designed for repeater operation…in favor of the greater good, particularly because many amateur repeaters are open to all amateur operators who desire to use them. The cooperation has taken the form of adherence to the determination of local frequency coordinators. While no amateur operator or amateur station 'owns' a frequency, this type of coordination is the minimum joint effort by the amateur community needed to facilitate repeater operation…

Several coordinators urged us to establish some mechanism to officially recognize local or regional coordinators. Others were concerned about the potential for abuse of power at the local level. Another concern was the exclusive right to coordinate within a geographical area. It is essential that repeater coordinators respond to the broadest base of local amateurs and consider the concerns not only of repeater owners, but also of those users of spectrum affected by repeater operation. Their authority is derived from the voluntary participation of the entire amateur community; their recognition must be derived from the same source. We believe the new rules will assure that a coordinator is representative of all local amateur operators.

Several commenters urged us to abolish closed repeaters in the amateur service or alternatively, to permit coordinators to relegate closed repeaters to secondary status or to give open repeaters preference when coordinating. We are not of the view, as were these commenters, that closed repeaters are any more or less desirable than open repeaters.

We proposed to make noncoordinated repeaters primarily responsible to resolve interference associated with coordinated repeaters. The ARRL commented that we should go further and make noncoordinated repeaters solely responsible to resolve such interference and require noncoordinated repeaters to cease operation if the interference is not resolved. Although the focus must be placed in the first instance upon the noncoordinated repeater to resolve such interference, we are adopting our proposed rules which continue to make the coordinated repeater secondarily responsible. This permits local coordinators and the FCC to consider technical alternatives, questions of equity, and spectrum efficiency in reaching the most reasonable solution.

This is the rationale behind the current Part 97 rules that affect amateur repeater coordination [97.201(c); 97.205(c)].

Band Plans

Band plan is a term you may encounter in the world of repeaters. It refers to an agreement between concerned VHF and UHF operators as to how each VHF and UHF Amateur Radio band should be used. The goal of a band plan is to systematically minimize interference among various modes sharing each band. Besides FM repeater and simplex activity, you can find CW, SSB, AM, OSCAR, digital, TV and R/C (radio control) operations, among others, on these bands.

Most hams who use or plan to use FM will probably use 2 meters. Check the description of the 2-meter band plan in Chapter 5.

In the 144.51-145.49 MHz range, the spacing between repeater channels is 20 kHz. The first input frequency is 144.51 MHz and the next is 144.53 MHz (144.51 MHz + 20 kHz = 144.53 MHz).

In the 146.01-147.99 MHz range, channel spacing is 20 or 30 kHz, depending on local policies.

In areas that use the 30-kHz spacing plan, some repeaters are on "splinter" (15-kHz spacing) frequencies, or between two repeater channels. For example, spacing would normally place repeater inputs on 146.01 and 146.04 MHz, but an additional repeater input can be established on 146.025 MHz.

According to the above band plan, the spacing between each repeater input and output is 600 kHz. For example, the input frequency of 146.34 MHz has an output frequency of 146.94 MHz (146.34 MHz + 600 kHz = 146.94 MHz).

The simplex frequencies between 144.90 and 145.10 MHz are generally used for packet operation, while those between 146.40 and 146.60 MHz, and between 147.40 and 147.60 MHz, are channelized for voice-simplex operation. (Select frequencies between 146.40 and 146.60 MHz and 147.40 and 147.60 MHz may also be repeater inputs or outputs.)

A common variation to the band plan is the reversal of the input and output frequencies of repeaters operating on splinter frequencies, such as 146.625 and 146.315 MHz. The purpose of this reversal is to reduce co-channel interference.

Under this band plan, repeater pairs on 146 MHz (i.e., 146.01/.61) have a low input/high output split; that is, the lower frequency (146.01) would be the input and the higher frequency (146.61) would be the output. Conversely, repeater pairs on 147 MHz (i.e., 147.81/.21) have a high input/low output.

Band plans for every amateur band can be found in Chapter 5. For information on band plans and repeater operating in your local area, contact your frequency coordinator, listed in the front of the ARRL *Repeater Directory*.

"Good Amateur Practice"

In an April 27, 1983, letter to a major repeater council, the FCC said:

The only national planning for Amateur Radio service frequencies that has come to our attention is that done by the American Radio Relay League. The 1982-83 edition of the ARRL *Repeater Directory* lists over 5600 stations in repeater operation all over the United States and Canada. In view of this widespread acceptance of their band plans, we conclude that any amateur who selects a station transmitting frequency not in harmony with those plans is not operating in accord with good amateur practice. For example, the ARRL

Repeater Directory lists the frequency pair 144.83/145.43 MHz as a repeater channel. Therefore, designation of this channel by the regional frequency coordinator in (an) area is in accord with the ARRL national band plan.

The bottom line is that if stations transmit on frequencies 144.83 or 145.43 (to use the Commission's example above) in a manner that creates interference to coordinated repeaters, sufficient cause would exist for issuance of an Official Notice of Violation. And such operation could mean an additional Notice for deliberate and malicious interference.

There are good reasons why simplex operation should be moved from repeater frequencies. The FCC has labeled three: (1) Repeater operation is not permitted on some Amateur Radio bands; (2) On bands where repeater operation is permitted, such operation is confined to a limited portion—a subband—of the band; and (3) The nature of repeater operation necessitates some form of channelization; haphazard frequency selection would result in poor spectrum use, and constant frequency change is impractical.

Simplex stations do not have such limitations; thus, they have much greater flexibility. They must avoid repeater channels in widely accepted band plans. Such avoidance constitutes good amateur practice. The FCC has fined amateurs operating simplex on repeater inputs for malicious interference [97.101(d)].

CHAPTER 7

Third-Party, Autopatch and Phone Patch Operation

C ommunications on behalf of third parties has been a part of Amateur Radio from its earliest beginnings in the United States. The very basis for the creation of the ARRL in 1914 was to organize amateurs to relay messages on one another's behalf in order to overcome the limited range of the amateur stations of the day. In that simpler time, there was essentially no limit to the content of such messages nor on whose behalf the messages could be sent.

It was not until the 1930s that international limitations were placed on amateur traffic, at the insistence of European governments for whom the telecommunications monopoly was a source of considerable revenue. While handling messages or providing communications for material compensation always has been prohibited in Amateur Radio, it was not until 1972 that the FCC specifically prohibited "business communications" in Docket 19245. Thus, the FCC began for the first time to regulate amateur traffic on the basis of its content.

Confusion developed on the part of the amateur community when it came to providing public service and meeting their personal communications needs. Hams had difficulty interpreting the rules, and in some cases, felt they were too restrictive. As a result, the Commission changed its rules once again. The focus of the new rules is not on content anymore, but on whether the amateur or his employer stand to gain financially. This has greatly simplified and expanded public service and personal communications opportunities for hams [97.113(a)].

What Exactly is Third-Party Traffic?

A third-party message is one the control operator (first party) of your station sends to the control operator of another station (second party) for anyone else (third party) [97.3(a)(44)]. Third-party messages include those that are spoken, written, keystroked, keyed, photographed or otherwise originated by or for a third party, and transmitted by your Amateur Radio station, live or delayed. A third party may also be a person permitted by the control operator to participate in Amateur Radio communications [97.115(b)].

There are three main types of third-party traffic. The first type involves *third-party messages,* those sent via traffic nets, packet networks, or other means. Message fairs and displays at shopping malls provide sources for this type of routine message. Not all messages need be as formal as this, however. A third-party message may be no more than a "Gee, while you're in Cheraw, say hi to Cousin John for me..."

The second type of third-party traffic involves *phone-line interconnection,* so-called "phone patch" and "autopatch" operation. This type of operation allows third parties to communicate directly with second and first parties via the telephone system.

The third type of third-party traffic is *direct participation* by interested third parties in actual Amateur Radio communications [97.115(b)]. The rules provide for this type of operation as long as the control operator continually monitors and supervises the third party's participation [97.115(b)(1)]. At no time may a control operator leave a third party unattended at a transmitter that is on the air. As long as the control operator is present and monitoring the communications, however, the third party may key the transmitter and identify the station; there is no requirement that the control operator actually do so. Also, the rules prohibit participation in this type of third-party traffic by a former amateur whose license was revoked; suspended for less than the balance of the license term and the suspension is still in effect; suspended for the balance of the license term and relicensing has not taken place; surrendered for cancellation following notice of revocation, suspension or monetary forfeiture proceedings; or who is the subject of a cease-and-desist-order that relates to amateur operation and is still in effect [97.115(b)(2)].

Third-party traffic involving payment for the communications service to any party is not allowed [97.113(a)(2)]. Third-party traffic in which the operator stands to gain financially or on behalf of the operator's employer is not permitted [97.113(a)(3)].

International Third-Party Traffic

International third-party traffic is traffic exchanged between control operators in different countries. It is prohibited for US operators except where:
- The US has a special third-party agreement with the other country (a list of these countries appears in the accompanying sidebar) [97.115(a)(2)] or;
- The third party is a licensed amateur [97.115(a)(2)] or;
- In cases of emergency where there is an immediate threat to lives and property [97.403].

Traffic carried over a message forwarding system must conform to the international regulations when the originating (first party) and receiving (second party) control operators are in different countries. If both first and second parties are in the same country, the international prohibition does not apply, even if the third party is in a different country. For example, if an amateur in California sends a message to an amateur in New York who then sends the message via the Internet to a third party in England, there is no violation because the over-the-air portion was within the US. In such a situation, the communication is delivered to or received from the third party via normal communications systems and thus there is no revenue loss to the foreign communications system.

At the end of any international third-party communications, a station must identify with its own call sign and that of the foreign station with which traffic was

Table 1

International Third-Party Traffic—Proceed With Caution

Occasionally, DX stations may ask you to pass a third-party message to a friend or relative in the States. This is all right as long as the US has signed an official third-party traffic agreement with that particular country, or the third party is a licensed amateur. The traffic must be noncommercial and of a personal, unimportant nature. During an emergency, the US State Department will often work out a special temporary agreement with the country involved. But in normal times, never handle traffic without first making sure it is legally permitted.

US Amateurs May Handle Third-Party Traffic With:

C5	The Gambia	J6	St Lucia	V7	Marshall
CE	Chile	J7	Dominica		Islands
CO	Cuba	J8	St Vincent and	VE	Canada
CP	Bolivia		the Grenadines	VK	Australia
CX	Uruguay	JY	Jordan	VR6**	Pitcairn Island
D6	Federal Islamic	LU	Argentina	XE	Mexico
	Republic of the	OA	Peru	YN	Nicaragua
	Comoros	PY	Brazil	YS	El Salvador
DU	Philippines	TG	Guatemala	YV	Venezuela
EL	Liberia	TI	Costa Rica	ZP	Paraguay
GB*	United	T9	Bosnia-	3DA0	Swaziland
	Kingdom		Herzegovina	4U1ITU	ITU Geneva
HC	Ecuador	V2	Antigua and	4U1VIC	VIC, Vienna
HH	Haiti		Barbuda	4X	Israel
HI	Dominican	V3	Belize	6Y	Jamaica
	Republic	V4	St Christopher	8R	Guyana
HK	Colombia		and Nevis	9G	Ghana
HP	Panama	V6	Federated	9L	Sierra Leone
HR	Honduras		States of	9Y	Trinidad and
J3	Grenada		Micronesia		Tobago

Notes

*Third-party traffic permitted between US amateurs and special-events stations in the United Kingdom having the prefix GB only, with the exception that GB3 stations are not included in this agreement.

** Since 1970, there has been an informal agreement between the United Kingdom and the US, permitting Pitcairn and US amateurs to exchange messages concerning medical emergencies, urgent need for equipment or supplies, and private or personal matters of island residents.

Please note that the Region 2 Division of the International Amateur Radio Union (IARU) has recommended that international traffic on the 20 and 15-meter bands be conducted on the following frequencies:
14.100-14.150 MHz
14.250-14.350 MHz
21.150-21.200 MHz
21.300-21.450 MHz
The IARU is the alliance of Amateur Radio societies from around the world; Region 2 comprises member-societies in North, South and Central America, and the Caribbean.
Note: At the end of an exchange of third-party traffic with a station located in a foreign country, an FCC-licensed amateur must transmit the call sign of the foreign station as well as his own call sign.

exchanged [97.115(c)]. The international Radio Regulations as well as the FCC's rules require that all international communications shall be made in plain language and must be of such an unimportant nature that recourse to the public telecommunications service is not justified [97.117].

Phone Patch and Autopatch Guidelines

Radio amateurs in the US enjoy a great privilege: The ability to interconnect their stations and repeaters with the public telephone system. The wisdom of the federal government in permitting, and even in defending, this freedom has been demonstrated time and again. There is no way to calculate the value of the lives and property that have been saved by the intelligent use of phone patch and autopatch facilities in emergency situations. The public interest has been well served by amateurs with interconnect capabilities.

As with any privilege, this one can be abused, and the penalty for abuse could be the loss of the privilege for all amateurs. What constitutes abuse of phone patch and autopatch privileges? In the absence of specific regulations governing their use, the answer to this question depends on one's perspective. Consider these facts: To other amateurs, phone patching activities that result in unnecessary frequency congestion or which appear as a commercialization of Amateur Radio operation are an abuse of their privilege to engage in other forms of amateur activities.

To the telephone company, which needs to protect its massive investment in capital equipment, anything that endangers its equipment, its personnel or its revenues is an abuse.

To the Federal Communications Commission, which is responsible for the efficient use of the radio spectrum by the services it regulates, any radiocommunication that could be handled more appropriately by wire is an unnecessary use of a valuable resource.

To the commercial suppliers of radiocommunication for business purposes (Radio Common Carriers), competition from a noncommercial service constitutes a possible threat to their livelihood.

At one time or another, threats to radio amateurs' interconnect privileges have come from each of these sources. And threats may come from another quarter: The governments of certain nations that prohibit amateurs from handling third-party messages internationally in competition with government-owned telecommunications services. If illegal phone patching to and from their countries cannot be controlled, they reason, the solution may be to ban all international third-party traffic by amateurs and to permit no such special arrangements.

The question facing amateurs is this: Should phone patches and autopatches be subject to reasonable voluntary restraints, thereby preserving most of our traditional flexibility, or should we risk forcing our government to define for us specifically what we can and cannot do? Experience has clearly shown that when specific regulations are established, innovation and flexibility are likely to suffer.

The Amateur Radio Service is not a common carrier, and its primary purpose is not the handling of routine messages on behalf of nonamateurs. Third-party communications as an incidental part of Amateur Radio, however, adds an important dimension to amateur public-service capability.

It is the policy of the American Radio Relay League to safeguard the prerogative of amateurs to interconnect their stations, including repeaters, to the public telephone system. An important element of this defense is encouraging amateurs to maintain a high standard of legal and ethical conduct in their patching activities. It is to this end that these guidelines are addressed. They are based on standards that have been in use for several years on a local or regional basis throughout the country. The ideas they represent have widespread support within the amateur community. All amateurs are urged to observe these standards carefully so amateurs' traditional freedom from government regulation may be preserved as much as possible.

1) International phone patches may be conducted only when there is a special third-party agreement between the countries concerned. The only exceptions are when the immediate safety of life or property is endangered, or where the third party is a licensed amateur.

2) Phone patches or autopatches involving the pecuniary interest of the originator, or on behalf of the originator's employer, must not be conducted at any time. The content of any patch should be such that it is clear to any listener that such communications are not involved. Particular caution must be observed in calling any business telephone. Calls to place an order for a commercial product may be made, but not calls to one's office to receive or to leave business messages. Calls made in the interests of highway safety, however, such as for the removal of injured persons from the scene of an accident or for the removal of a disabled vehicle from a hazardous location, are permitted.

3) All interconnections must be made in accordance with telephone company tariffs. This means that your equipment must not affect the proper functioning of the telephone system; if it does, you are responsible for correcting the problem. If you have trouble obtaining information about the tariffs from your telephone company, they are available for public inspection at the telephone company office.

4) Phone patches and autopatches should never be made solely to avoid telephone toll charges. Phone patches and autopatches should never be made when normal telephone service could just as easily be used.

5) Third parties should not be retransmitted until the responsible control operator has explained the nature of Amateur Radio to them. Control of the station must never be relinquished to an unlicensed person. Permitting a person you don't know very well to conduct a patch in a language you don't understand amounts to relinquishing control.

6) Make sure the third parties know they are participating in radio communications, and that such communications are not private, and may be heard by people other than the parties involved.

7) Phone patches and autopatches must be terminated immediately in the event of any illegality or impropriety.

8) Autopatch facilities must not be used for broadcasting. If a repeater can transmit information of general interest, such as weather reports, such transmissions must occur only when requested by a licensed amateur and must not conform to a specific time schedule. The retransmission of radio signals from other services except for NOAA weather, government propagation bulletins and space shuttle communications, is not permitted in the Amateur Radio service.

9) Station identification must be strictly observed. In particular, US stations conducting international phone patches must identify in English at least once every 10 minutes, and must give their call signs and the other stations' call signs at the end of the communication.

10) In selecting frequencies for phone patch work, the rights of other amateurs

must be considered. In particular, patching on 20 meters should be confined to the upper portion of the 14,250-14,350 kHz segment, in accordance with the IARU Region 2 recommendation.

11) Phone patches and autopatches should be kept as brief as possible, as a courtesy to other amateurs; the amateur bands are intended to be used primarily for communication among radio amateurs.

12) If you have any doubt as to the legality or advisability of a patch, don't make it.

Compliance with these guidelines will help ensure that hams' interconnection privilege will continue to be available in the future, which will in turn help amateurs contribute to the public interest.

Reverse Autopatch

No unlicensed person may initiate an amateur transmission without the knowledge and approval of the station's control operator. Incoming calls to an autopatch must be answered and screened off the air by the control operator to ensure rule compliance. If an incoming call automatically causes the repeater to transmit, even if it's just a signal tone or notification message, then it is possible for an unlicensed person to initiate a transmission without the control operator's knowledge or approval, which is not permitted. The use of a reverse autopatch is permitted only under very limited conditions.

The Simplex Autopatch

A simplex autopatch operates very much like a repeater autopatch, except the station it is attached to is not a repeater. The station is operating on a simplex frequency and does not automatically retransmit the signals of other amateur stations. The main use for a "simpatch" is to provide a mobile user with the same functionality as a repeater patch from his home station.

The most important point to remember is that a control operator must be present at both stations' control points [97.105(a); 97.115(b)(1)]. The operation involves third-party traffic in that a third party is "participating" in amateur radiocommunication [97.115(b)]. This "participation" is permitted as long as each station involved has a control operator monitoring and supervising the radiocommunication continuously to ensure compliance with the rules [97.115(b)(1)]; automatic control is not permitted [97.109(e)]. There is no problem if a control operator is present at the fixed station location. If, however, the mobile user wishes to act as the control operator of the fixed station, then a radio link must be used to remotely control the fixed station, and that link is an auxiliary station [97.213(a)]. The link therefore must be on frequencies permitted for auxiliary station operation [97.201(d)]. A simplex autopatch may not be controlled via a 2-meter link. See Chapter 6.

CHAPTER 8
Specialized Operating

Space Operations

When you build a repeater, you try to put the antenna on a tall tower or on top of a tall building. The higher you get the antenna, the better coverage your repeater will have. What about a repeater with an antenna more than 50 kilometers high? An amateur satellite is a repeater in space with a large "footprint" on the Earth. Except for geostationary orbits, satellites don't stay in one spot, either. They cross international borders all the time. A satellite is not your average amateur station.

In 1971, member nations of the International Telecommunication Union (ITU) sat down at the conference table to fashion regulations for the burgeoning commercial and noncommercial satellite technologies. Rules were needed for international coordination of satellite frequencies, locations and purposes, given the proliferation of hardware floating virtually over every nation on Earth. The 1971 Space WARC (World Administrative Radio Conference) delegates arrived at a number of agreements and requirements, and the FCC was obligated to carry them out with regard to the services it regulates. Accordingly, a new subpart (H) dealing specifically with the Amateur Satellite Service was introduced into Part 97. In 1989 these rules were integrated into Subpart C (Special Operations). As with other sections of the rules, the FCC has taken steps to bring the space operations section of Part 97 into close accord with the ITU regulations.

Space Stations

The FCC defines a *Space Station* as "An amateur station located more than 50 km above the Earth's surface" [97.3(a)(38)]. This includes amateur satellites, amateur operation from the space shuttle, the Russian *Mir* space station and any future operations by astronauts in space. Special rules apply to space stations:

• Any station can be a space station. Any licensed ham can be the control operator of a space station [97.207(a)]. Control operator privileges are subject to the limitations of the control operator's license, of course. If you have a Technician class license, you can't control a satellite on a 15-meter uplink outside the 15-meter Novice/Technician Plus subband.

• It must be possible to make the space station stop transmitting if the FCC orders this [97.207(b)]. This is done by telecommand, another type of specialized operation examined later in this chapter. The FCC wants to make sure someone can

Table 1

Frequencies Authorized to Space Stations [97.207(c)]

Bands Wavelength	Frequency
40 m	7.0-7.1 MHz
20 m	14.00-14.25 MHz
17 m	18.068-18.168 MHz
15 m	21.00-21.45 MHz
12 m	24.89-24.99 MHz
10 m	28.0-29.7 MHz
2 m	144-146 MHz
70 cm	435-438 MHz
23 cm	1260-1270 MHz
13 cm	2400-2450 MHz
9 cm	3.40-3.41 GHz in ITU Regions 2 and 3 only
5 cm	5.83-5.85 GHz
3 cm	10.45-10.50 GHz
1.2 cm	24.00-24.05 GHz
6 mm	47.0-47.2 GHz
4 mm	75.5-81.0 GHz
2 mm	142-149 GHz
1 mm	241-250 GHz

shut off OSCAR-29's transmitter if it's interfering with the guidance system on the Jupiter 2 as it leaves for Alpha Centauri.

• A space station may automatically retransmit the radio signals of Earth stations and other space stations. An amateur satellite wouldn't be very useful without this provision [97.207(d)].

• Space stations are exempt from some rules that apply to ordinary amateur stations. Space stations do not have to identify themselves, and are permitted to transmit one-way communications [97.207(e)].

• If space stations transmit telemetry (the results of measurements made in the station) [97.3(a)(39)] they may use special codes to facilitate communications without worrying about the "codes and ciphers" provisions in Part 97 [97.207(f)].

• Space stations may transmit only on certain frequencies authorized in Part 97. Table 1 shows the transmit frequencies authorized to space stations [97.207(c)].

• The licensee of a space station must notify the FCC. This is an important part of placing an amateur satellite in orbit. Notification allows the FCC to keep track of US-licensed amateur satellites so that it can respond to any interference complaints registered with the ITU. A satellite is an "international" station and its licensee must remember that he has international responsibilities. The FCC specifies when they must be notified in Part 97, but rather than spell out what information must be provided, the FCC defers to the ITU Regulations. The appropriate sections of the ITU Regulations are available from ARRL HQ. If you see a space station in your future, these rules are required reading!

The FCC must be notified four times [97.207(g),(h),(i)]:

1) No less than 27 months before the space station will begin transmissions. This notification must include the information specified in Appendix 4 of the ITU Regulations. This includes the name of the satellite, the date it will commence

operations, the name and address of the organization responsible for the satellite and the satellite's orbital parameters (where it will travel in space). Notification must include technical information about any radio links the satellite will use; Earth-to-space, space-to-Earth and space-to-space as appropriate [97.207(g)(1)].

2) The FCC must be notified again no less than five months before the space station starts transmitting. This time Appendix 3 is the relevant part of the ITU Regulations. This notification must include more detailed information about the RF links the space station will use [97.207(g)(2)].

3) Once the space station is in space and operating, you have seven days to let the FCC know your station has commenced in-space operation. This notice must update the pre-space notifications. For example, you must tell the FCC if the 23-cm transponder on your satellite has failed or the 13-cm telemetry transmitter is only operating at half power [97.207(h)].

4) Finally, when the satellite fails or the shuttle lands and space operation is terminated, you must notify the FCC within three months. If the FCC orders you to terminate space operation, you must let them know that you have complied no later than 24 hours after the operation ends [97.207(i)].

Earth Stations

The FCC defines an *Earth Station* as "an amateur station located on, or within 50 km of, the Earth's surface intended for communications with space stations or with other Earth stations by means of one or more objects in space" [97.3(a)(15)]. Any radio amateur can be the control operator on an Earth station, subject to the limitations of his license class [97.209(a)]. Specific transmitting frequencies are authorized for Earth stations [97.209(b)]; these are shown in Table 2.

Table 2
Frequencies Authorized to Earth and Telecommand Stations [97.209 and 97.211]

Bands

Wavelength	Frequency
40 m	7.0-7.1 MHz
20 m	14.00-14.25 MHz
17 m	18.068-18.168 MHz
15 m	21.00-21.45 MHz
12 m	24.89-24.99 MHz
10 m	28.0-29.7 MHz
2 m	144-146 MHz
70 cm	435-438 MHz
23 cm	1260-1270 MHz
13 cm	2400-2450 MHz
9 cm	3.40-3.41 GHz in ITU Regions 2 and 3 only
5 cm	5.65-5.67 GHz
3 cm	10.45-10.50 GHz
1.2 cm	24.00-24.05 GHz
6 mm	47.0-47.2 GHz
4 mm	75.5-81.0 GHz
2 mm	142-149 GHz
1 mm	241-250 GHz

Space Telecommand Stations

A *Telecommand Station* "transmits communications to initiate, modify or terminate functions of a space station" [97.3(a)(42)]. A few special rules apply to a telecommand station:

• Any amateur station designated by the licensee of a space station is eligible to transmit as a space telecommand station to the space station. This is a simple precaution that protects the space station from unauthorized control. This privilege is limited by the privileges of the operator's license class [97.211(a)]. Like the space station control operator requirements, if you have a Technician class license, you can't control a space telecommand station on 20 meters.

• A space telecommand station may transmit special codes intended to obscure the meaning of telecommand messages [97.211(b)]. This protects the space station from unauthorized control commands. One wrong command could turn an expensive satellite into just a piece of orbiting space junk!

• Space telecommand stations may transmit only on certain frequencies [97.211(c)]. Table 3 shows the frequencies authorized for telecommand stations.

Table 3

Frequencies Authorized for Automatically Controlled Beacon Stations [97.203(d)]

28.20-28.30 MHz
50.06-50.08 MHz
144.275-144.300 MHz
222.05-222.06 MHz
432.300-432.400 MHz
all amateur bands above 450 MHz

Telecommand of Model Craft

We mentioned remote control (*telecommand*) of an amateur station [97.213] in Chapter 6. Amateurs are also permitted to use radio links to control "model craft," such as model airplanes and boats. Certain restrictions apply [97.215] on this kind of operation:

• Station identification is not required for transmission directed only to the model craft. The control transmitter must bear a label indicating the station's call sign and the licensee's name and address [97.215(a)].

• Control signals are not considered codes and ciphers [97.215(b)].

• Transmitter power cannot exceed 1 W [97.215(c)].

• Only licensed amateurs may operate telecommand transmitters using amateur frequencies. While unlicensed persons may participate as "third-parties" in most amateur operations, they may not participate in telecommand operation. This is true even when a licensed amateur is closely supervising the operation, or when a two operator "buddy box" is being used. The FCC has said that the one-way transmissions involved in telecommand do not constitute third-party messages exchanged between control operators. Nonamateurs must use equipment and frequencies in the Radio Control Service. If licensed amateurs wish to use amateur equipment and frequencies for model telecommand, it is their responsibility to be sure that they can safely operate the equipment while observing the FCC's Rules.

Beacons

A *beacon* is simply a transmitter that alerts listeners to its presence. In the Radionavigation Service, beacons are used to provide navigational guidance. In the Amateur Service, beacons are used primarily for the study of radio-wave propagation—to allow amateurs to tell when a band is open to different parts of the

KD4JL and KC4NEQ work on the AMSAT Phase 3D satellite, scheduled to be launched by an Ariane rocket in April 1996. Amateurs worldwide are contributing to this effort. *(photo courtesy Ed Richter, KD4JL and Rod Davis, KC4NEQ)*

country or world. Accordingly, the FCC defines a beacon as "an amateur station transmitting communications for the purposes of observation of propagation and reception or other related experimental activities" [97.3(a)(9)].

The rules address beacon operation [97.203]. A few key points:

• Automatically controlled beacon stations are limited to the frequencies shown in Table 3. Beacons that are manually controlled are not subject to the same restrictions as automatically controlled beacons [97.203(d)].

• The transmitter power of a beacon must not exceed 100 W [97.203(c)].

• Any license class, except Novice, can operate a beacon station [97.203(a)].

There is one exception to the rules laid out in §97.203. In 1979 the FCC issued a special license for an automatically controlled beacon operating only on 14.100 MHz. This beacon is part of the Northern California DX Foundation/IARU sponsored worldwide system of beacons on that frequency.

Spread Spectrum

People think of an "efficient" communications system as one that crams the most information into the smallest possible spectrum space. Hams have been taught for years that "wide" means "bad" as far as radio signals go. Most of the operations detailed in this chapter are relegated to limited frequency bands. *Spread-spectrum* (SS) techniques take the opposite approach: The signal is spread out over a very wide bandwidth.

The ARRL *Handbook* contains a detailed technical description of spread-spectrum techniques, so we will not go into much technical detail here. It is sufficient to say that SS uses digital techniques to spread a signal out over a wide bandwidth in such a way that the signals can be recovered by a receiver expecting the SS signal. The signal may be spread so far that it becomes invisible to a conventional receiver. This has obvious military applications. Another benefit of the SS system is that the "despreading" process can reject undesired signals—even signals much stronger than the desired communications. This makes SS a popular military anti-jamming technique.

SS communications are permitted on the 70-cm (420-MHz) band and all amateur frequencies above this band [97.305(c)]. Amateurs are allowed to use two types of SS techniques. Frequency hopping (FH) spread spectrum uses a binary sequence to switch the transmit frequency many times per second. If the receiver is synchronized to the transmitter, it can follow the hopping and the system becomes transparent to the users. To avoid interference to and from conventional users of a band, the time the transmitter stays on any one frequency must be kept short. This time (called the dwell time) is typically less than 10 milliseconds [97.311(c)(1)].

Direct sequence (DS) spread spectrum uses a fast bit stream to spread the energy of the carrier over a wide band of frequencies. This binary sequence is called pseudonoise (PN). The transmitter and receiver must use the same synchronized sequence, but once they are in sync, the system becomes transparent to the user [97.311(c)(2)].

The FCC Rules for spread-spectrum operation are fairly complicated and intimidating at first. As you read the rules, however, you'll find that they merely spell out the parameters that must be followed to ensure that everyone uses the same kind of SS. The rules were written to make it easy for the FCC to monitor this new mode. The FCC also hopes to minimize interference complaints and to make the few that do arise easier to deal with. Let's look at a few highlights of the rules:

• SS communications are authorized only between points within areas regulated by the FCC. This means that under the current rules, you can't use SS to work your buddy in Canada, even if Canadian hams are permitted to use SS [97.311(a)].

• SS communications must not be used for the purposes of obscuring the meaning of any communication. An obvious provision, but one FCC wanted to spell out, given the potential of SS techniques [97.311(a)].

• Stations transmitting SS communications must not cause interference to other stations, and must accept interference from stations employing other authorized emissions. This says that for now, SS is a "second-class citizen" on the ham bands. Your complaints that the local repeater is interfering with your SS communications will fall on deaf ears at the FCC. On the other hand, if your SS communications are interfering with the Orange Tabby Swap'n Shop Net, you will be told to cease and desist. The FCC states in this provision that unintentionally triggering a carrier-operated repeater is not considered harmful interference. It's still not a good idea, however. If you cause this kind of interference when operating frequency-hopping SS, back off on your dwell time on any particular frequency. Tests in the Washington, DC, area showed that fast FHSS with short dwell times can operate on the inputs of repeaters without "kerchunking" or causing noticeable interference in the outputs of the repeaters [97.311(b)].

• Only certain spreading sequences may be used. This allows the FCC to monitor SS communications [97.311(c)].

• Detailed records of SS operations must be kept [97.311(e)]. These records must be kept for one year following the last entry and must contain enough information to allow the FCC to demodulate your SS signals.

• SS transmitter power must not exceed 100 W [97.311(g)].

These rules may seem a bit excessive, but they're there to allow experimentation while protecting other users of the ham bands where SS is permitted. With time, the FCC may relax these restrictions and allow SS to take its place with the other "first-class" emissions.

Summary

Part 97 allows unconventional operating modes and techniques. Widen your horizons and experiment with them, but remember that you are responsible for understanding the rules that pertain to anything you do on the air. It's an old cliché, but "ignorance of the law is no excuse." Read the appropriate Sections of Part 97 until you are sure that you are following the letter and the spirit of the rules. If you have questions, the Regulatory Information Branch at ARRL HQ is there to help.

Summary

CHAPTER 9

Antennas and Restrictions

Radio amateurs consider an antenna a thing of beauty, the essential conduit to the exciting world of Amateur Radio communication. Amateurs have been erecting antennas and support structures (such as towers, poles and the like) for nearly a century. During this time, local municipalities have adopted zoning ordinances and regulations as part of their duty to preserve property values and what the law calls the public health, safety, convenience, and welfare. Beauty is in the eye of the beholder, of course, whether it be a five-element monobander atop a 65-foot tower, a large billboard sign advertising a local bistro, groundbreaking for another shopping mall or an addition on your home. All these matters are affected by zoning. You shouldn't get the impression that hams have been *solely* singled out for attention by local politicians. The role of local regulators, however, has often proved incompatible with the legitimate need for radio amateurs to put up the most effective antennas possible within their individual circumstances.

Restrictions imposed on amateur antennas come in more than one flavor: besides public zoning ordinances, nongovernmental private deed restrictions and covenants specific to a particular parcel of land are common. Don't buy the beverages and pizza for your antenna-raising party unless you've done your homework. Before we delve into the critically important world of local zoning ordinances and private covenants, a word about the few antenna restrictions imposed by Uncle Sam is in order.

Federal Restrictions on Antennas

Fortunately, there are few restrictions on antennas in Part 97. Those that are listed only apply in fairly unusual circumstances.

Antennas Higher Than 200 Feet

You must have the FCC's approval before you put up a tower more than 200 feet high [97.15(a)]. Use Form 854 to notify the FCC and Form 7460-1 to notify the Federal Aviation Administration (FAA) of your proposed tower [97.15(d)]. If you must file these forms, you will also have to mark and light your antenna. See Part 17, *Construction, Marking and Lighting of Antenna Structures*, of the FCC Rules. You may request a copy from the Regulatory Information Branch at ARRL HQ.

Federal Communications Commission

Request for Approval of Proposed Amateur Radio Antenna and Notification of Action

Approved by OMB
3060-0139
Expires 3/31/89

Instructions:

- Complete this form only if the height of the antenna will exceed –
  60.96 meters (200 feet) or –
  1/100 of the minimum distance between the antenna site and any aircraft landing area.
- Mail this application to the FCC, Antenna Survey Branch, Washington, D.C. 20554.
- See Statements on back.

FCC USE ONLY

Approved by:

Date:

☐ No obstruction marking required.

☐ Obstruction marking required.

☐ Form 715 par. ____

☐ Form 715A par. ____

1. First Name   M.I.   Last Name

2. Mailing Address: Number & Street   Telephone No. ( )

   City   State   ZIP Code   3. Amateur Call Sign

4. Antenna Location

   NOTE: To find the north latitude, west longitude, and site elevation of the antenna location, use a 7.5 minute topographic quadrangle map. This may be purchased from the U.S. Geological Survey, Washington. D.C. 20242 or U.S. Geological Survey, Denver, Colorado 80225. You must determine your north latitude and west longitude to the nearest second.

	Degrees	Minutes	Seconds		Degrees	Minutes	Seconds
North Latitude				West Longitude			

5. Check the type of support structure proposed for the antenna. ☐ Tower ☐ Pole ☐ Bldg. ☐ Other (specify) ____

6. (a) Give the elevation of the ground above mean sea level at antenna site . . . ____ feet
   (b) Give the overall height of the antenna structure above ground level . . . . . ____ feet
   (c) The overall height above mean sea level is (add 6(a) and 6(b)) . . . . ____ feet

7. Will this antenna require notification of proposed construction to the FAA? ☐ Yes ☐ No (See FCC Rules & Regulations Section 17.7 & 17.14) If "Yes" give the following information:

Location of FAA Office Notified	Date of Notification	FAA Study Number

8. Will the antenna share an existing structure? ☐ Yes ☐ No If "Yes" give the following information:

Name of Licensee	FCC Call Sign	Radio Service

I certify that I am the above-named applicant and that all statements made on this application and any attachments hereto are true and complete to the best of my knowledge.

Signature   Date   A willfully false statement is a criminal offense; U.S. Code, Title 18, Section 1001

FCC USE ONLY

When validated by the FCC Seal, this form and any attachment(s) become part of your amateur radio station license. Keep it with your station files.
NOTE: THIS IS NOT AN AUTHORIZATION TO TRANSMIT.

Licensee's Name	Amateur Call Sign	Location of Antenna
		City State

Antenna Coordinates

North Latitude	West Longitude
Degrees - Minutes - Seconds	Degrees - Minutes - Seconds

Remarks:

NOT VALID WITHOUT FCC SEAL

☐ Antenna height approved at ____ feet (AGL)

☐ No obstruction marking or lighting

☐ Marked & Lighted per FCC Form(s):

☐ 715 Par: ____

☐ 715A Par: ____

FCC 854
May 1986

Fig 1—FCC Form 854

Environmental Impact Statement

If you have filled out a Form 610 lately to renew/modify your license or to sit for an exam, you've noticed that Item 6 asks if your application would have a significant environmental impact; virtually everyone answers "no." You would only answer "yes" in those rare instances where you plan to put up an antenna in an officially designated wilderness area, or on land that is significant in American history, architecture or

HOW AND WHERE TO NOTIFY FAA

Forward one completed set of FAA Form 7460-1 to the Chief, Air Traffic Division, Federal Aviation Administration, (Name of office - see below), for the office having jurisdiction over the area within which the proposed construction or alteration will be located. The geographic area of jurisdiction for each FAA office is indicated below:

FAA REGIONAL BOUNDRIES
Includes Locations of Regional Headquarters and Centers

ADDRESSES OF THE FAA REGIONAL AREA OFFICES

AAL - Alaskan Region

Alaskan Regional Office
Air Traffic Division
701 "C" Street
Anchorage, AK 99513
Mail Address
701 "C" Street Box 14
Anchorage, AK 99513
Tel. (907) 271-5892

ANM - Northwest Mountain Region

Northwest Mountain Regional Office
Air Traffic Division
17900 Pacific Hwy South
C-68966
Seattle, WA 98168
Tel. (206) 431-2530

AGL - Great Lakes Region

Great Lakes Regional Office
Air Traffic Division
2300 East Devon Avenue
Des Plaines, IL 60018
Tel. (312) 694-7458

ACE - Central Region

Central Regional Office
Air Traffic Division
601 East 12th Street
Kansas City, MO 64106
Tel. (816) 374-3408

AEA - Eastern Region

Eastern Regional Office
Air Traffic Division
JFK International Airport
Federal Building
Jamaica, NY 11430
Tel. (718) 917-1228

ANE - New England Region

New England Regional Office
Air Traffic Division
12 New England Executive Park
Burlington, MA 01803
Tel. (617) 273-7150

AWP - Western-Pacific Region

Western-Pacific Regional Office
Air Traffic Division
15000 Aviation Boulevard
Hawthorne, CA 90260
Mail Address
P.O. Box 92007
Worldway Postal Center
Los Angeles, CA 90009
Tel. (213) 297-1182

ASW - Southwest Region

Southwest Regional Office
Air Traffic Division
4400 Blue Mound Road
Fort Worth, TX 76101
Mail Address
P.O. Box 1689
Fort Worth, TX 76101
Tel. (817) 624-4911 Ext. 306

ASO - Southern Region

Southern Regional Office
Air Traffic Division
3400 Whipple Street
East Point, GA 30344
Mail Address
P.O. Box 20636
Atlanta, GA 30320
Tel. (404) 763-7646

FCC 854
May 1986

culture, or where extensive changes in surface features are required, or where the tower must be lighted and is located in a residential area [97.13(a)]. If you do answer "yes," you need to submit an environmental assessment with your Form 610; see §§1.1301-1.1319 of the FCC Rules (available from HQ) for more details.

Antennas Near Airports

The FCC specifies maximum allowable heights for antennas located near air-

Form Approved OMB No. 2120-0001

NOTICE OF PROPOSED CONSTRUCTION OR ALTERATION

US Department of Transportation
Federal Aviation Administration

Aeronautical Study Number

1. Nature of Proposal

A. Type	B. Class	C. Work Schedule Dates
☐ New Construction	☐ Permanent	Beginning _____
☐ Alteration	☐ Temporary (Duration _____ months)	End _____

2. Complete Description of Structure

A. Include effective radiated power and assigned frequency of all existing, proposed or modified AM, FM, or TV broadcast stations utilizing this structure.

B. Include size and configuration of power transmission lines and their supporting towers in the vicinity of FAA facilities and public airports.

C. Include information showing site orientation, dimensions, and construction materials of the proposed structure.

3A. Name and address of individual, company, corporation, etc. proposing the construction or alteration. *(Number, Street, City, State and Zip Code)*

( )
area code   Telephone Number

B. Name, address and telephone number of proponent's representative if different than 3 above.

(if more space is required, continue on a separate sheet.)

4. Location of Structure

A. Coordinates *(To nearest second)*	B. Nearest City or Town, and State	C. Name of nearest airport, heliport, flightpark, or seaplane base
° ' " Latitude		(1) Distance from structure to nearest point of nearest runway — Miles
° ' " Longitude	(1) Distance to 4B	(2) Direction from structure to airport
	(2) Direction to 4B	

5. Height and Elevation *(Complete to the nearest foot)*

A. Elevation of site above mean sea level

B. Height of Structure including all appurtenances and lighting (if any) above ground, or water if so situated

C. Overall height above mean sea level (A - B)

D. Description of location of site with respect to highways, streets, airports, prominent terrain features, existing structures, etc. Attach a U.S. Geological Survey quadrangle map or equivalent showing the relationship of construction site to nearest airport(s). (if more space is required, continue on a separate sheet of paper and attach to this notice.)

Notice is required by Part 77 of the Federal Aviation Regulations (14 C.F.R. Part 77) pursuant to Section 1101 of the Federal Aviation Act of 1958, as amended (49 U.S.C. 1101). Persons who knowingly and willingly violate the Notice requirements of Part 77 are subject to a fine (criminal penalty) of not more than $500 for the first offense and not more than $2,000 for subsequent offenses, pursuant to Section 902(a) of the Federal Aviation Act of 1958, as amended (49 U.S.C. 1472(a)).

I HEREBY CERTIFY that all of the above statements made by me are true, complete, and correct to the best of my knowledge. In addition, I agree to obstruction mark and/or light the structure in accordance with established marking & lighting standards if necessary.

Date	Typed Name/Title of Person Filing Notice	Signature

FOR FAA USE ONLY — *FAA will either return this form or issue a separate acknowledgement.*

The Proposal:

☐ Does not require a notice to FAA.

☐ Is not identified as an obstruction under any standard of FAR, Part 77, Subpart C, and would not be a hazard to air navigation.

☐ Is identified as an obstruction under the standards of FAR, Part 77, Subpart C, but would not be a hazard to air navigation.

☐ Should be obstruction ☐ marked, ☐ lighted per FAA Advisory Circular 70/7460-1, Chapter(s) _____

☐ Obstruction marking and lighting are not necessary.

Remarks:

Supplemental Notice of Construction FAA Form 7460-2 is required any time the project is abandoned, or

☐ At least 48 hours before the start of construction.

☐ Within five days after the construction reaches its greatest height.

This determination expires on _____ unless:

(a) extended, revised or terminated by the issuing office;
(b) the construction is subject to the licensing authority of the Federal Communications Commission and an application for a construction permit is made to the FCC on or before the above expiration date. In such case the determination expires on the date prescribed by the FCC for completion of construction, or on the date the FCC denies the application.

NOTE: Request for extension of the effective period of this determination must be postmarked or delivered to the issuing office at least 15 days prior to the expiration date.

If the structure is subject to the licensing authority of the FCC, a copy of this determination will be sent to that Agency.

Issued In	Signature	Date

FAA Form 7460-1 (8-85)

Fig 2—FAA Form 7460-1

ports. The airport must be:

1) available for public use and listed in the *Airport Directory* of the current *Airman's Information Manual* or in the *Alaska* or *Pacific Airman's Guide and Chart Supplement*, or

2) an airport or heliport under construction that is the subject of a notice or proposal on file with the FAA, and except for military airports, it is clearly indi-

cated that the airport will be available for public use; or

3) an airport or heliport operated by the US armed forces.

If you have one of these airports near you, then the following limits apply to you:

If the runway is longer than 1 km (3280 feet) and the airport is within 6.1 km (3.79 miles) of your proposed installation, your antenna may be no higher than 1 meter (3.28 feet) above the airport elevation for every 100 meters (328 feet) from the nearest runway. This is a slope of 100 to 1. See Figure 3.

If the runway is shorter than 1 km (3280 feet) and the airport is within 6.1 km (3.79 miles) of your proposed installation, your antenna may be no higher than 2 meters (6.56 feet) above the airport elevation for every 100 meters (328 feet) from the nearest runway. This is a slope of 50 to 1. See Figure 4.

If the installation is within 1.5 km (4920 feet) of a helipad, your antenna may be no higher than 4 meters (13.1 feet) above the airport elevation for every 100 meters (328 feet) from the nearest landing pad. That's a slope of 25 to 1. See Figure 5.

For example, your antenna tower, a mile away from a runway longer than 3280 feet, must be within the slope of 100 to 1 if you want to avoid the hassle of

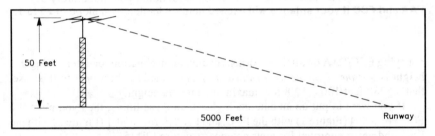

Fig 3—If you live near an airport runway more than 3280 feet long, you must notify the FAA and FCC if your antenna will exceed heights limited by a slope of 100 to 1.

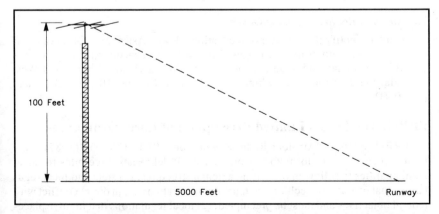

Fig 4—Where airport runways are less than 3280 feet long, you must notify the FAA and FCC if your antenna will exceed heights limited by a slope of 50 to 1.

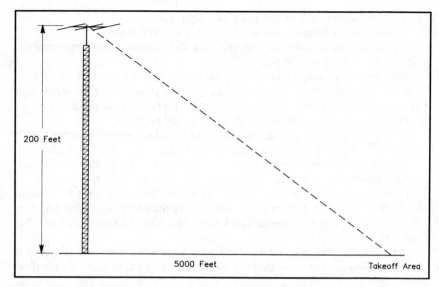

Fig 5—If there is a heliport near your antenna location, you must notify the FAA and FCC if your antenna will exceed heights limited by a slope of 25 to 1.

Within the figure:
- 200 Feet
- 5000 Feet
- Takeoff Area

notifying FCC/FAA on all those forms. A convenient equation for determining this height is *distance from nearest runway in feet* divided by the *slope*. In this case, that's 5280 feet/100 = 52.8 feet maximum antenna height.

If you want to put up an antenna higher than these limits, then you must file FCC Form 854 (Figure 1) with the FCC, and FAA Form 7460-1 (Figure 2) with the FAA and await approval for your antenna structure [97.15].

If your antenna will not exceed 20 feet above the ground or 20 feet above the natural or man-made features at its site, it is exempt from the height restrictions described above. The roof of your house counts as a man-made structure if your antenna is mounted on it, but antenna towers do not [97.15(c)].

The National Radio Astronomy Observatory

You must notify the Director of the National Radio Astronomy Observatory if you intend to put up or make changes to a repeater or beacon antenna in what is called the National Radio Quiet Zone, an area in Maryland, Virginia and West Virginia, the coordinates of which are defined in Chapter 10 [97.203(e)] and [97.205(f)].

PRB-1, The FCC's Limited Preemption of Local Ordinances

PRB-1, cited as "Amateur Radio Preemption, 101 FCC2d 952 (1985)," is a limited preemption of local zoning ordinances. It delineates three rules for local municipalities to follow in regulating antenna structures: (1) state and local regulations that operate to preclude amateur communications are in direct conflict with federal objectives and must be preempted; (2) local regulations that involve placement, screening or height of antennas based on health, safety or aesthetic considerations must be crafted to reasonably accommodate amateur communications;

and (3) such local regulations must represent the minimum practicable regulation to accomplish the local authority's legitimate purpose. The heart of PRB-1 is codified in the FCC Rules [97.15(e)]. Of course, what is "reasonable" depends on the circumstances! For suggestions about what to do when you are faced with a restrictive ordinance, see "Interacting with Municipal Officials," below. The full text of PRB-1 can be found in Appendix 7. A package containing PRB-1 and supplementary materials may be obtained by sending a 9″ × 12″ SASE with $3.00 (14 oz.) of postage to the Regulatory Information Branch at ARRL HQ.

Q&A—PRB-1

Q. How can the federal government limit the local zoning power of the mayor, city council or zoning board, who are much closer to local land-use conditions and the community?

A. PRB-1 recognizes that local leaders can regulate amateur installations to ensure the safety and health of people in the community. This general function has been termed part of a state and local government's "police powers" by legal authorities. But PRB-1, in striking a balance between the proper exercise by the city of its police powers *and* the federal interest in promoting the Amateur Radio service, holds that such regulations cannot be so restrictive that amateur communications are impossible. Nor can the regulations be more restrictive than necessary to accomplish the purpose of protecting the community.

The theory of PRB-1 and its incorporation into the rules [97.15(e)] is that there is a reasonable accommodation to be made between the amateur's communications needs and the obligation of the zoning authorities to protect the community's health, safety and general welfare. Just as it is the zoning official's legitimate concern, for example, that your tower doesn't come crashing down on your neighbor's house, it is the FCC's legitimate concern (as delegated by Congress) that you be allowed to legally put up a tower of reasonable size so that you can successfully conduct interstate and international radio communications on the ham bands.

Q. Why doesn't PRB-1 or Section 97.15(e) specify a "reasonable" height?

A. Whole legal textbooks have been devoted to trying to define what's "reasonable." As discussed above, the line between legitimate amateur communication goals and the local authority's interests must be determined on a case-by-case basis at the local level. The FCC declined to indicate what it considered a reasonable height below which a city cannot regulate antennas. What might be reasonable in one area might not be reasonable in others. For example, erecting a full-size 80-meter Yagi atop a 150-foot tower behind a townhouse in a densely populated suburb probably would not be considered reasonable. Yet the same antenna might be entirely acceptable in a rural environment. As a general principle, however, a municipality that establishes a blanket height limitation of any type, especially one that doesn't permit at least 65 feet of antenna height, will have a difficult time justifying that limitation as a technical matter in light of PRB-1.

Q. Now that PRB-1 and Section 97.15(e) have put the brakes on local zoning authorities, can I just put my antennas up to what I consider a reasonable height and assume that the ordinance in my city is in violation of the FCC's official policy?

A. *Absolutely not!* The PRB-1 Order and its codification in 97.15(e) is no more than a statement of policy by the FCC. It's not a panacea; it's not an overnight cure by any means. If you believe your city's ordinance is not valid in light of PRB-1, it's up to *you* to establish that and get the ordinance changed by the city council or have the existing ordinance declared invalid by the courts. To violate the ordinance or to put up an antenna without a permit can subject you to serious fines and even criminal penalties. In encouraging local municipalities to pay appropriate attention to the federal interest in amateur communications, the FCC said that if you believe the local zoning authorities have been overreaching in terms of interfering with your ability to operate Amateur Radio, you can use the PRB-1 document and the FCC rule [97.15(e)] to bring FCC policies to their attention. The bottom line is that it's still up to you to *prove* that the existing ordinance is not in accordance with PRB-1.

Q. What do I need to do to prove that the ordinance in my city is in violation of PRB-1?

A. You should be able to establish that you need an antenna of a certain height to reliably communicate on HF, VHF and/or UHF, as the case may be. Your city is usually concerned with aesthetics, safety, property values and RFI. As to the interrelated issues of aesthetics and property values, the city's interests must be balanced against yours. You probably can't convince your neighbor that a three-element tribander looks "good" if she feels otherwise. It can, however, be established as to what effect, if any, an antenna of a certain height will have on property values. The written opinion of a local professional appraiser can be a crucial piece of evidence for your side. The safety factor can best be addressed through exhibition of tower manufacturer's specifications for proper installation. Explain that there is no relationship between antenna height and safety. The safety issue is best dealt with by ensuring the integrity of your installation, especially relative to the size of the base section and proper guying. As to RFI, you might assert (although interference is a sensitive subject to your neighbors) that regulation of RFI is an FCC matter and not appropriate for local zoning regulation (see discussion in Chaper 10). But hasten to add that your modern equipment has less chance of causing interference to television sets than a neighbor's hair dryer or cordless baby monitor.

Materials are available from ARRL HQ to assist you in this regard. Send a large (9 × 12) SASE with $3.00 (14 oz.) postage to the Regulatory Information Branch at HQ to obtain a "PRB-1 Package." HQ can also refer you to ARRL Volunteer Counsels and Volunteer Consulting Engineers in your area. These are ham attorneys and structural engineers willing to provide an initial consultation about your zoning matter free of charge.

Q. What if I can't get any satisfaction from the building inspector and/or the Zoning Board of Appeals, and I decide to take the city to court. Has PRB-1 been used in any court cases successfully?

A. Yes. John Thernes, WM4T, successfully sued the city of Lakeside Park, Kentucky, after he initially received an adverse determination by a federal district court judge. Thernes had applied in 1982 for a building permit for a 78-foot tower and antennas. The city denied his application, claiming that the zoning ordinance did not permit antennas and support structures. Thernes sued the city, but a US District Court dismissed his complaint, noting that there was no statement from the FCC preempting local zoning ordinances. Thernes appealed to the US Court of Appeals for the Sixth Federal Circuit (which covers Kentucky) and on the eve of oral argument in that case, the FCC released PRB-1. That was enough for the Appeals Court to send the case back to the District Court for reconsideration.

The same district judge who had ruled against Thernes previously indicated this time around that because of PRB-1, he was inclined to rule *in favor of* Thernes. By agreement of the parties, judgment was entered against the city and in favor of Thernes, and he was finally able to put up a 73-foot tower. He received an award of attorney's fees, as well. The judge, in the consent decree, noted that PRB-1 obligates municipalities to cooperatively arrive at an accommodation for amateur antennas in local zoning ordinances [The final disposition of this case is cited at 62 Pike and Fisher Radio Regulations 2d, 284 (1987)].

Another US District Court Judge used PRB-1 to void an antenna ordinance in Sands Point, New York, which limited antennas to no more than 25 feet. The case had been brought by noted contester Andrew Bodony, K2LE, to overturn the denial by Sands Point officials of his application for a building permit and later a variance (which was also denied) for an 86-foot crank-up tower. The court found PRB-1 to be a proper exercise of FCC authority and that the town's blanket 25-foot height limitation seriously interfered with the ability of Bodony to operate his Amateur Radio station. The court also noted that there was no showing that Bodony's tower would endanger the health, safety or general welfare of neighborhood residents or was detrimental to the character of the neighborhood [The judge's decision is cited at 61 Federal Supplement 2d, 307 (1987)].

You should consider taking your matter to court only if you've exhausted all your options and appeals with your town government. You must observe all the legal technicalities, including deadlines, on your way to your final administrative appeal. Only then do you have grounds to take your matter to the court system. Make sure the record at the zoning board level is as complete as possible (include every bit of evidence you can muster) because it is generally that record, that paper trail, that a court will rely on most heavily. If you then take your case to court, be prepared for the long haul; you must have the time, money and patience to see it through.

Q. What if I avoid all the hassle and just put up my antenna, hoping no one will notice?

A. This question is often asked because we all know of cases where one ham simply put up an antenna and didn't have any trouble, while another one in the same town, following the rules like a good citizen, applied for a building permit for his tower and had all kinds of grief from zoning authorities and/or his immediate neighbors. Remember, all it takes is one complaint to cause problems not only for you, but possibly all the other radio amateurs in town, if you haven't secured the proper permit when one is required. A ham who has no permit is in a difficult spot if he or she is found to be in violation of the building code or zoning ordinance.

Q. Does PRB-1 mean that I can ignore the "no-antenna" clause in my apartment lease or condo bylaws?

A. *No.* In PRB-1, the FCC addressed zoning ordinances only; it exercised no federal preemption over restrictive covenants in private contractual agreements (such as your apartment lease). It explained that becaue these agreements are *voluntarily* entered into by the buyer or tenant and seller or landlord by contract, rather than established by the government, they don't come under FCC jurisdiction (at least for the time being). The ARRL has asked the FCC to reconsider this. The subject of covenants and deed restrictions is taken up later in this chapter.

Local Zoning Ordinances

In the past, amateurs relied solely on their powers of persuasion when dealing with local officials. Conflicts between amateurs and local authorities over the antenna height, placement in the yard, number of antennas on a particular support structure (eg, a tower) and the like were common. In the absence of detailed federal regulations governing amateur antennas (except for those aspects discussed previously), municipal leaders often fill in the void and use their broad discretion in public health and safety matters to enact regulations that limit antennas and supporting structures. The people who write these regulations have a lot of other things on their mind, so these regulations seldom take into account your need for an antenna of certain dimensions and height to be effective (working the DXpeditions, running phone patches to the South Pole and so on), so conflicts arise.

The situation reached epidemic proportions in the early 1980s and amateurs who invested family savings in fighting local zoning, building codes and covenant restrictions in the courts around the country were losing because there was no clear statement of any federal interest in the matter by the FCC. The courts held that the FCC regulates radio, but because the FCC had issued no statement restraining the zoning power of cities and counties, the traditionally local interest in zoning regulations that protects the public generally superseded the interests of any individual amateur.

By October 1983, the ARRL Board of Directors reviewed the adverse court decisions and recognized that antenna restrictions would continue to be a major stumbling block unless a statement of federal preemption emerged from the FCC. On July 16, 1984, the League filed a formal request asking the FCC to issue a declaratory ruling that would declare void all local ordinances that preclude or significantly inhibit effective, reliable amateur communications. Hundreds of comments were filed when the FCC established a pleading cycle, labeled PRB-1 ("PRB" being the designation for the FCC's Private Radio Bureau, the bureau in the FCC's internal organization that handled Amateur Radio matters at that time. It has been replaced by the Wireless Telecommunications Bureau). Comments were filed by amateurs, zoning authorities and city planners.

September 19, 1985, was a red-letter day in the history of Amateur Radio, as the FCC issued its now-famous PRB-1 declaratory *Memorandum Opinion and Order*, which says, in pertinent part, that "state and local regulations that operate to preclude amateur communications in their communities are in direct conflict with federal objectives and must be preempted." See Appendix 7 for the complete text of PRB-1.

May 31, 1989, marked another milestone when the Commission adopted the revised and reorganized Part 97. The new rules codify the essence of the PRB-1 ruling: ". . . State or local regulation of amateur antennas may not preclude, but must reasonably accommodate, such communications, and must constitute the minimum practicable regulation to accomplish the local authority's legitimate purpose" [97.15(e)].

The specific holding of PRB-1 has been of extreme benefit to amateurs and, with a few exceptions, has encouraged open cooperation and dialogue between the communities seeking to regulate amateur antennas and amateurs. Now that important language of PRB-1 has been incorporated into the FCC Rules, the federal interest and official FCC policy with respect to amateur communications can be more easily demonstrated to municipal officials who need to be educated by you and your fellow hams.

Interaction with Municipal Officials

Don't be intimidated by the thought of going to city hall for a permit. Virtually all ham radio operators who own the physical area necessary for the safe installation of a tower should be able to legally erect a tower of *some* size. Here are the steps to take to enhance your chances of getting as much tower as you wish:

Information Gathering

Because regulations pertaining to antennas and the way building and zoning departments (or the equivalent) process permits vary from city to city, the first and most important step is information gathering. This means a visit or a phone call to your local building/zoning department or the equivalent, to obtain a copy of the zoning ordinances. Don't settle for anything less than the *whole booklet* of regulations! If the clerk or secretary offers to photocopy for you only the pages that pertain to antennas, keep in mind the sections mentioned, and politely thank him or her. But then get the entire booklet (which normally costs between $5 and $15).

You should get the whole booklet because you need to know more than just the sections under which the town regulates antenna heights. You need to know what procedures to follow when you apply for a permit. You also need to know how to appeal an adverse decision if you don't get a favorable ruling from the building inspector (or the zoning enforcement officer, or the equivalent) on the first try. Furthermore, if you ever need to seek the advice of a lawyer, the first thing the attorney will need to see is the entire body of regulatory law affecting land-use regulations for the town. Therefore, obtain the entire book and study the regulations carefully.

Zoning Regs Defined

But what are zoning regulations, exactly? Zoning regulations are rules that establish the permitted uses and the minimum and maximum dimensional requirements of structures in established areas or "zones." Ninety-nine percent of the time, a ham will want to put up an antenna/tower at his home, which will be in a residential zone. Because the overwhelming number of jurisdictions hold Amateur Radio to be a normal, accessory (as opposed to primary) use of residential property, it is proper in a residential zone. Similarly, such edifices as swimming pools, tennis courts and tool sheds are considered accessory structures on residential property.

But in addition to use rules, zoning regulations also establish rules as to how *high* structures are permitted to be. You may find your proposed tower being held to the same height standards as other "buildings" or "structures" permitted in a residential zone. Read the definitions section near the beginning of the zoning rules. Sometimes the definition of "building" is broad enough to include a tower, or antennas may be defined specifically. If the regulations do *not* define antennas specifically, see if they mention "accessory structures." Read the definition to see if antennas are included in that definition. You may find a section that defines flagpoles, church steeples and similar structures in language that could easily apply to towers as well.

Your town may be concerned with building codes, which are standards relating to safety that have been agreed upon by engineers from the architectural, structural, civil and other engineering disciplines. Once the building inspector determines that your tower is proper and would not violate zoning regulations pertaining to use and dimensions, your construction must still be carried out in accordance with building codes. Fortunately, this is rarely a problem. Tower manufacturers provide detailed specifications and plans for proper installation in accordance with all building codes.

Meeting the Building Inspector

After you have had a chance to study the regulations, you can probably tell which rules apply to your installation. If the clerk or secretary pointed out certain sections, look at those sections first to see if you agree. The regulations, when read

Obviously "The Old Man," Hiram Percy Maxim (1AW), didn't have zoning problems when he erected this antenna at his home in Hartford, Connecticut in the 1920s.

in the context of your proposed antenna installation, should be understandable. If the regulations are full of legal mumbo-jumbo, however, now is the time to consult a lawyer, such as an ARRL Volunteer Counsel. But if you feel confident about your level of understanding and have familiarized yourself enough to carry on an intelligent conversation about the regulations, make an appointment with the building inspector (or the appropriate city official). Be prepared to discuss the proposed location, height and purpose of your structure, and take along the basic engineering data provided by the tower manufacturer to satisfy building code concerns. Also, take along a rough drawing of your property that shows your boundary lines, the house and other buildings nearby, and the proposed location of the tower.

The building inspector, much like a police officer walking a beat, is the first interpreter of the law, in this case the zoning law. What he or she says will be the first indication of the steps you will have to take to get a permit for your installation. After you present your proposal, listen carefully to what the building inspector says. Building inspectors are often willing to be helpful and grant your permit, provided you follow the correct application procedures. For example, you may need to file a map of your property drawn to scale.

If the building inspector appears negative, pay attention nevertheless to what he has to say, even if his reasoning may be wrong. It is important to thoroughly understand the basis for his opinion. Do not go into the confrontation mode with the building inspector; keep your "grid current" low! Don't wave a copy of PRB-1 or Part 97 in his face and "command" him to give you the permit because of federal law. He's not going to know what you're talking about, nor is he likely to make a snap decision in your favor that might get him into hot water later. If you are going to talk legal issues, moreover, he's going to want his lawyer (the town attorney) in on it.

In many situations, the building inspector will have the authority to grant a building permit or other approvals without the involvement of any of the higher-ups. This is why it's important to maintain a good relationship with him if possible.

Sometimes (especially if you're lucky), zoning ordinances specifically exempt antennas from the height restrictions of other structures. Depending on how your town's ordinances are written, it may be necessary for you to seek the permission of a zoning commissioner or land-use board. Usually this means that the zoning regulations are set up to allow antennas to a certain height limit without the need of a hearing, but if you want to exceed the "usual" height threshold, you need to apply for a special permit. This means that the drafters of the zoning ordinances decided that certain uses of structures could only be permitted after a public hearing and demonstration of special need.

While this undoubtedly means more red tape and delay if you find yourself in this situation, such a requirement is not illegal in the eyes of the law (including PRB-1). It does provide a forum for potential opposition from neighbors, however, and preparation for the hearing is all-important.

If there is a problem, it will be one of two varieties: It may be a matter of interpretation of the ordinance by the building inspector, or the ordinance may be written in such a way that he is unable to reasonably come up with any other interpretation. If the ordinance is prohibitive under any reasonable interpretation, you should immediately seek the advice of a lawyer.

If the problem is that you disagree with the building inspector's interpretation of the zoning regulations, that's not as serious. While you may ultimately need a lawyer to resolve the issue, you can still carry the issue further yourself. Tell the building inspector in a *nonconfrontational* manner that you have a different interpretation and ask him for his comments. See if you can narrow it down as to where the problem lies. If it is a problem with the way the ordinance is written, ie, under no possible interpretation can you get your permit, that is a more serious problem. It may mean that the ordinance is illegal, and therefore invalid. The town officials won't like that and are more apt to fight vigorously against having their ordinance invalidated. They would rather change an interpretation for a particular situation than throw out their entire ordinance and have to start from scratch. If you run into an unresolvable problem with the building inspector, you should then, as a last resort, calmly tell him about PRB-1 and the rules [97.15(e)]. As mentioned previously, more often than not, you will find that it does not help at the building inspector level. At most, all you may get across is that the federal government, through the FCC, has acted under a preemption order called PRB-1. You can say that local governments cannot prohibit antenna towers, nor can they unreasonably restrict them in terms of size and height. However, you can't expect the building inspector to be equipped to engage in a lengthy discussion of the nuances of the legalities of federal preemption.

The Appeal

If there is a reasonable interpretation or even a loophole in the regulations under which you should be allowed to put up the antenna/tower, you can usually appeal that decision to your town's Zoning Board of Appeals (ZBA) or equivalent body. If you have absolutely no alternative, you can ask the ZBA for a variance. This means you are asking the ZBA to relax the zoning rules in your case, to give you a special exemption because of some exceptional difficulty or unreasonable hardship. Variances are granted sparingly because it is difficult to establish severe hardship. Remember, too, that if you apply for a variance, you are admitting in the eyes of the law that the ordinance applies to your antenna. This means that you can't contest or challenge the applicability or the jurisdiction of the ordinance in a later proceeding.

The procedure for appealing the building inspector's ruling is outlined in the regulations. You are given a chance to explain what Amateur Radio is and to present your alternative interpretation that would permit you to legally erect your tower or to explain that your hardship is severe enough that you ought to be allowed to put up your tower despite the zoning rules.

PRB-1 should be used to persuade the ZBA to adopt your more reasonable interpretation because federal law requires it to adopt as reasonable an interpretation as possible. If there is no possible interpretation of the zoning regulations that will allow your tower to go up, then tell the ZBA that PRB-1 and Section 97.15(e) are binding federal regulations that supersede the ZBA's law if there is a conflict. That is, its interpretation of its own zoning ordinance should be guided by the binding order rendered by the FCC in PRB-1. In other words, the local regulations cannot regulate in the "overkill mode." But make no mistake about it; state and local governments can, under the specific language of PRB-1, still regulate antennas for reasons of health, safety and welfare, as long as the regulations are reasonable.

Additional Guidelines for the Presentation

Here are a few other things to keep in mind when you make your presentation before the ZBA: Make sure you can establish the safety factors of your tower. On matters of safety, there should be *no compromise* by the municipality. If your installation does not meet building-code requirements, no board (or court, if it comes to that) will allow it. The manufacturer's specifications must be followed.

On the other hand, aesthetics and welfare, particularly the effect on the surrounding property values, are more likely to be areas of compromise by the ZBA. Your task here is threefold: You have to demonstrate your *need* for the proposed tower and the *safety* of the structure, and you have to show that you have taken reasonable steps to *lessen the impact on the surrounding property.*

As mentioned previously, ARRL HQ has helpful materials to assist you in preparing for the hearing in the "PRB-1 Package." The League can also refer you to ARRL Volunteer Counsels or Consulting Engineers near you. Contact the Regulatory Information Branch at ARRL HQ for details.

If you haven't done so, try to gauge the opposition, if any, of your neighbors prior to the hearing. Neighbors who show up at the hearing can be friendly or adversarial; touching base with your neighbors in advance is a good way to transform adversaries into people who will speak in support of your position. Be prepared to answer all questions that you can anticipate. Although the touchy area of radio-frequency interference (RFI) does not fall under the jurisdiction of the ZBA, your reaction to questions about RFI might be used by the ZBA members to judge your character (which could form the unwritten or unstated basis for a denial). A ZBA member is more likely to give the benefit of the doubt to someone who sounds like a responsible, good neighbor.

The touchy subject of RFI came up earlier in this chapter, in the discussion of PRB-1. If neighbors or ZBA members raise concerns about RFI, one of the most effective statements you can make is to explain that, although years ago the home-brew nature of ham stations may have resulted in hams being responsible for RFI, today that is hardly the case. With the present level of Amateur Radio sophistication, RFI is rarely a problem, and when it is, it is usually the fault of the manufacturers of the stereo, TV or other home-entertainment device. But explain to the ZBA members that you will work with your neighbors to resolve any RFI problems in the rare event that RFI problems occur. Also, point out that a tower taller than neighbors' homes will help direct your signals *above* their houses, providing added "insurance" against possible problems.

When you make your presentation, keep it clear, concise and simple. Use a written outline so you don't forget key points, but speak directly and respectfully to the ZBA members in your own words. Avoid technical language or ham radio jargon that's incomprehensible to non-hams. The ZBA does not, and has no reason to, care about dB, SWR, wavelengths, DX or anything else "sacred" to Amateur Radio operators. They care about how big your aerial is going to be and what damage it could do to a neighbor's home if it falls (you might point out that towers rarely fall or break, and if they do, it is generally not from the bottom). Emphasize the public service nature of Amateur Radio, and its value to your community in an emergency.

Make sure your demeanor is professional, and dress in a conservative business-like manner (leave your ARES jumpsuit, call-sign cap, and painter's pants at home). Make sure, also, not to operate in a vacuum; enlist the support of the local Amateur Radio community and ask as many of your fellow hams as possible to show up for the hearing. A unified show of support will make a profound impression on ZBA members who are, after all, ordinary people yet politicians at heart.

Private Restrictions

There are circumstances under which it is legally *impossible* to erect a tower. Condominium owners are a prime example. A condo owner "owns" only that which exists *within* the confines of the four walls that forms his unit. The rest of the building and the land are owned by someone else or owned in common by all the unit owners. Unless you can persuade the condominium association to allow you to put up an antenna in a common area, you had better concentrate on operating mobile from your car or bone up on so-called "invisible" or limited-space antennas in attics or crawl spaces in the ceiling (although this may also be a violation of the condominium bylaws). You cannot expect to be able to put up a tower on land or a building that you do not own outright. Private restrictions, commonly called

CC&Rs (covenants, conditions and restrictions) are another aspect of antenna/ tower regulation that exist outside of the zoning regulatory body. CC&Rs are the "fine print" that may be referenced in the deed to your property, *especially* if you are in a planned subdivision that has underground utilities. Your best bet before proceeding with your tower plans is to take a copy of your deed to an attorney to have a limited title search done for the specific purpose of determining whether private restrictions will affect your rights. If you're buying a new home, have a title search done *before* you sign on the dotted line. When you agree to purchase the property, you agree to accept covenants and deed restrictions on the land records, which may preclude you from being able to put up a tower at your new dream home. See **Deed Restrictions** below.

Summary

With respect to governmental, as opposed to private, restrictions, it all boils down to one simple fact: Any licensed ham radio operator who has the private property of a sufficiently sized parcel of land, has the right under federal law to erect a tower and antenna, subject to the reasonable regulation of local and state government.

A substantial amount of money has been spent in legal costs by amateurs in attempts to have courts tell local governments that their idea of reasonable is *not* reasonable under PRB-1. Sometimes a clash cannot be avoided. But if you're like most of us, you want to know how best to go about putting up the highest tower you can without getting into a big legal struggle and alienating your otherwise friendly neighbors.

Emphasis from the start should be on the practical aspects of planning a successful campaign for getting the necessary approvals:

• Be realistic about the physical limitations of your backyard.

• Go on an information-gathering mission at the local town hall to determine which officials and regulations are applicable to your proposed antenna installation.

• Discuss your plans with the local building inspector.

• Determine whether the building inspector's interpretation of how the zoning rules apply to your plans is fair or should be respectfully challenged.

• Then, if you are dissatisfied with the inspector's interpretation, obtain information from ARRL HQ and seek legal advice from a competent attorney as to whether there is any reasonable interpretation of the regulations that would allow you to put up the tower.

• If no such interpretation presents itself, seek a variance or make a frontal assault on the ordinance as being in violation of PRB-1. Once you get into this area of the law, legal advice is often a necessity.

Most amateurs who follow these steps come out of it with a positive result. If a final decision by your local zoning board is unsatisfactory, however, you have the option of taking the matter to court.

Deed Restrictions

Long before zoning regulation of land existed, private restrictions in deeds controlled how land could be used. The English system of common law, that we have inherited, permitted a seller of land to impose certain restrictions on the use of that land, which the seller, even after the sale was long past, could enforce in the courts. These restrictions or covenants were included in the deed from seller to buyer. Today, as noted above, deed restrictions are typically referred to as covenants, conditions and restrictions, *CC&Rs*.

These days, covenants are commonly used, especially in new housing developments, by builders or developers as a means of controlling land use after individual parcels are sold off. Suppose a builder has a tract of 100 homes and is selling them one at a time. The builder, until he is able to sell all of the lots, wants to maintain uniformity so that, for example, one of the buyers cannot paint his house 20 different day-glo colors, making the remaining unsold lots less marketable.

Since the CB boom of the mid-1970s, builders and developers have included antennas in the standard list of things they don't want homeowners to install while the builder is attempting to sell the remainder of the houses in the development. They are concerned that people may find antennas or towers unattractive. So when the subdivision plan is filed in a town's land records, a list of covenants is filed, too. Every deed from the builder makes reference to the list, subjecting every buyer down the line to the restrictions.

Typically, the declaration of covenants provides that enforcement authority passes from the developer to the homeowner's association after the developer sells all the houses in the development and goes on his merry way. The homeowner's association is, among other things, charged with maintaining the aesthetics of the neighborhood and can often determine whether additional structures, such as a tool shed, swing set or antenna can be built. If a homeowner installs an antenna in violation of a covenant, the homeowner's association can bring that owner to court to enforce the covenant.

ARRL has a covenant package available which explains the nature of covenants. It also contains suggested language which can be incorporated into a purchase or

---

Q&A—Deed Restrictions

Q. How can I find out if there are any deed restrictions on my land?

A. Deed restrictions and/or covenants are filed with the town clerk (or the equivalent) in your town (or county, depending on your state). These records are completely open to the public; you don't have to be a lawyer or a real estate broker to examine land records. Some restrictions are mentioned in the deed you received at your closing, so check your paperwork. Such restrictions need not be mentioned in your individual deed, however. A master list of covenants, usually termed "Declaration of Covenants" or something similar, is a list of things that cannot be done on a particular tract of land and may be referred to in subsequent deeds from the seller to the buyer. The land is sold to a buyer "subject to" the covenants on file. Even if you do not buy your land from the original builder who filed them, covenants are considered to "run with the land"; they are referenced in the deed to any particular buyer. Even though you may never have seen the list of covenants on file and even though they are not fully explained in your particular deed, and yes, even though you did not buy your home directly from the builder who established them, you are nevertheless bound by them. Why? Because you voluntarily bought land that is itself subject to those restrictions, and as such, in the eyes of the law (remember, justice is blind!), you have submitted to those covenants and agreed to them. This is why attorneys spend hours looking through boring, old land records; they're trying to find the "land mines" before it's too late. Before you sign a contract to buy a new home, have your attorney or other competent title searcher check the land records for covenants.

Q. How can I avoid antenna covenants when I buy my new house?

A. You can't, but you may be able to avoid surprises. Make sure your contract for the purchase of land specifies that the sale is contingent on the absence of any deed restrictions, covenants or conditions that would prevent or restrict your ability to install an amateur antenna at least (fill in the blank) feet in height. Require that within two weeks of the signing of the sales contact, the seller must give you a notarized statement to that effect. If it turns out that there are covenants, you could attempt to obtain approval for your proposed antenna from the homeowner's association, architectural control committee or the builder, in accordance with the covenants, before signing a contract to buy the house.

Don't rely on oral representations of anyone involved. Unless you have written, unconditional authorization from whoever has the enforcement authority, don't assume that you will be able to put up your tower after you move in. Your attorney should be consulted before signing anything.

Q. In my area, I can't even buy a new or recently built house without finding restrictions in the deed. Can't I use PRB-1 to invalidate them?

A. Sorry, no. The courts and the FCC view antenna covenants as a private contractual matter between a buyer and seller. This may not be the most realistic view today, because more and more amateurs have few alternatives in their home buying because of the pervasiveness of covenants. It may be hard to find real estate in your area without covenants. Unfortunately, the fact that land without covenants is unavailable in a particular area does not make a good legal challenge to deed restrictions.

Q. Some amateurs have successfully asserted their rights to put up a tower despite deed restrictions prohibiting them. How can this be done?

A. There are some standard defenses to covenants. The best one is to show that the restriction has not been enforced against others in the neighborhood. If there are CB antennas or other amateur antennas, it may not be possible to enforce the restriction against you. Even in this case, however, there are limitations. One court has held that the existence of TV antennas in a development where covenants prohibited all antennas did not constitute "abandonment or waiver" of the covenant. An amateur could not put up a beam because that was held to be significantly different from small TV antennas. Also, don't assume that abandonment of one covenant will extend to others. If, for example, your neighbor keeps a boat in his driveway in violation of a covenant, this would not waive an anti-antenna covenant.

There are other defenses, however: The longer a covenant goes unenforced, the more difficult it is to enforce, as courts require prompt objections to covenant violations. Further, if actions of the homeowner's association have led the amateur to believe that approval was given or that the covenants would not be enforced, you can argue that you should be allowed to keep your antenna in the air, even in violation of a covenant.

Q. The covenants in my development do not prohibit antennas or towers, but do require the approval of the homeowner's association. How do I convince them to approve my antenna?

A. Tactics differ, but generally the best strategy is to prepare in advance for the questions you are likely to encounter: (1) How you have planned the installation to minimize visual impact; (2) How it will be safe; (3) How it will enable you to conduct emergency and public service communications for the neighborhood; (4) How it will *not* affect property values; (5) How RFI is minimized by higher antennas; and (6) That you need it to communicate effectively and reliably. Also, follow the general guidelines for the ZBA presentation discussed previously, as applicable.

The best thing to do is sit down with your immediate neighbors on all sides, explain in a friendly way what you'd like to do, and assure them that the Voice of America is not being constructed next door to them! If you can obtain the approval of your immediate neighbors, the homeowner's association is less likely to object. If you anticipate opposition, consider the advantages of a crank-up tower and as a compromise, offer to keep it cranked down when you're not using it.

Q. It sounds as though the FCC, in PRB-1, has protected us from some antenna restrictions imposed by my town, but not from my neighbors. What gives?

A. The ARRL Board of Directors considers covenants the most serious cumulative problem that faces amateurs at present. The League has submitted data to the FCC demonstrating that deed restrictions are not simply a matter of contractual agreement between the buyer and seller. The League continues to work on this problem; stay tuned.

lease agreement. To get it send an SASE with 2 units of First Class postage to the Regulatory Information Branch at ARRL HQ.

Conclusion

Some hams are blessed with living in communities where there are neither zoning nor covenant restrictions on their antenna farms. However, you can't assume that this is the case in your own particular situation. The time to start looking into whether it's okay to put up a tower is *before*, not *after*, you start pouring the concrete for your base section. If you exercise common sense (ie, resist the temptation to put up a Big Bertha on a city lot) and good-faith compliance with the legal procedures, you will undoubtedly be successful in securing permission for an appropriately sized tower for your antenna system. But keep in mind that what you do, and how you do it, affects not only you, but all other radio amateurs in your town, and perhaps—through precedent-setting court cases—amateurs throughout the entire country.

CHAPTER 10

Interference: EMI/RFI

Radio Frequency Interference

While both modern amateur transmitters and consumer electronics equipment are less prone to interference than was equipment made several decades ago, the complexity of electronic equipment has grown considerably over the years. The net result is that hams and their neighbors can still experience *electromagnetic interference* (EMI) problems when a radio transmitter is operated near consumer electronics equipment. Consumer electronics equipment can cause interference to hams, as well. Nearly every ham has been chased off the radio when a family member turns on an electric motor or video game. It would take an entire book to cover the subject in detail; in fact, the ARRL book, *Radio Frequency Interference: How to Find It and Fix It* has chapters that cover nearly all technical and regulatory aspects of interference problems. You can order it from ARRL Headquarters or buy it at most ham-equipment dealers.

Most new hams learn about interference when a neighbor comes a-knockin' to complain about interference to consumer electronics equipment. Your neighbor will not understand all of the technical and regulatory details about interference; all he or she knows is that when you are on the air, their equipment doesn't work correctly, so it must be all your fault. While this may be true, it is more likely that their consumer electronics was not designed to work near a radio transmitter and may require additional filtering and shielding to function properly when you are on the air.

Dealing with the interpersonal aspects of interference problems usually must happen before it is possible to deal with the technical and regulatory aspects. Each person in such a case has a unique perspective on the situation, and a different degree of understanding of the technical and regulatory issues involved. On the other hand, each person also has certain responsibilities to the other party and should be prepared to address those responsibilities fairly.

Interference has two basic causes: a radio transmitter (or noise source) can be transmitting signals that are outside of its allocated frequency. These signals are called *spurious emissions* or *spurs*. If one of your transmitter spurs is inside the frequency used by a nearby TV station, it is likely that you will cause interference to many TVs in your neighborhood. Interference from spurious emissions can only be fixed at the transmitter.

On the other hand, consumer electronics equipment should be able to select only the desired radio signal, without "hearing" other radio signals. As an example, if you were to tune your TV to channel 4, and you saw channel 9 instead, you would probably decide that something was wrong with that TV. If you tune your TV to channel 4, and are picking up a local HF or VHF transmitter instead, the TV may require additional filtering and shielding to function near a radio transmitter. This is called *fundamental overload*. Fundamental overload can only be fixed at the affected equipment.

There are no regulations regarding the RF susceptibility of consumer electronics equipment. Manufacturers have developed voluntary standards of immunity, and modern equipment is actually better than equipment designed and built a few years ago. However, not all manufacturers adhere to these standards and they are under no legal obligation to help consumers with interference problems.

It is important that you and your neighbor understand the difference between the two basic types of interference. To help you, the ARRL has prepared a pamphlet that explains interference in nontechnical terms. Send a self-addressed, stamped envelope to the ARRL Technical Department Secretary at ARRL Headquarters. Ask for the *What to Do if You Have an Electrical Interference Problem* pamphlet.

To help minimize interference between hams and other radio services, the FCC has written several rules about the "spectral purity" of radio transmitters, see Chapter 12. The most important rule as far as EMI is concerned are the absolute limits on spurious emissions that fall outside the amateur bands. These are described in detail in §97.307, and shown in graph form in Figs 2 and 3 in Chapter 12. The level of spurious signals varies from band to band, and with power level, but here are some examples that cover typical amateur operation: For typical HF QRP operation, all spurious emissions must be at least 30 dB less than the fundamental signal; for HF "barefoot" operation, spurious emissions must be attenuated at 40 dB; for HF 1500-W operation, spurious signals must be attenuated 44 dB; and for VHF operation, the required attenuation ranges from 46 dB for 1-W transmitters to 60 dB for 25-W operation.

The levels defined in §97.307 are absolute-maximum levels of spurious emissions. This means that your spurious emissions must not exceed these levels, *whether the spurious emissions are causing interference or not.* Nearly all modern amateur equipment meets the requirements spelled out in the rules.

In addition, FCC regulations state that if a spurious emission from a transmitter causes interference to another radio service, the operator of the transmitter must take whatever steps are necessary to reduce or eliminate the interference. At the edge of a TV station's range, you might need 70 or 80 dB of attenuation of your spurious emissions, to ensure that a harmonic doesn't cause interference with a relatively weak TV signal. (This doesn't apply to "fringe" reception. If your neighbor is trying to receive a TV signal from 200 miles away, the FCC does not offer any protection to a TV signal that is that weak.)

Meeting these regulations is all that Part 97 *requires* you to do. If interference

FCC Interference Handbook

In 1995, the FCC released a new *Interference Handbook*, which you can get from your local FCC Field Office (listed in Appendix 12). The book takes a fair and honest approach to explaining responsibilities and cures for interference problems; it's a fine treatment of this complex technical and emotional subject, as shown by the following extracts:

FROM PAGE 1:

Many interference problems are the direct result of poor equipment installation. Cost-cutting manufacturing techniques, such as insufficient shielding or inadequate filtering, may also cause your equipment to react to a nearby radio transmitter. This is not the fault of the transmitter and little can be done to the transmitter to correct the problem.

FROM SECTION G, TELEPHONE EQUIPMENT INTERFERENCE:

Telephone interference generally happens because telephones are not designed to operate near radio transmitters and the telephone improperly operates as a radio receiver.

FROM SECTION I, INTERFERENCE TO OTHER EQUIPMENT:

Stereos, electronic organs, and intercom devices, among others, can react to nearby radio transmitters. When this happens, the device improperly functions as a radio receiver. You should first determine what type of interference you are receiving. See Part II of this bulletin for assistance. You may try to relocate the device within your home. It is usually impractical or expensive to modify the affected device. Contact the manufacturer for assistance or consider changing to another brand or model.

---

to consumer equipment is caused by fundamental overload, it cannot be fixed at your transmitter; it must be fixed by adding filtering or shielding to the affected equipment. In cases involving non-radio equipment, the FCC's opinion is that these devices should not pick up radio signals at all.

Quiet Hours

If you are causing interference, the FCC can impose so-called "quiet hours" on your station's operation. These quiet hours, when you would not be permitted to operate, are usually from 8 PM to 10:30 PM local time daily and on Sunday from 10:30 AM to 1 PM.

In almost all cases, the FCC only applies quiet hours to a station that has been demonstrated to be causing interference from spurious emissions. The quiet hours are applicable only to situations where there is interference with reception of transmissions of the domestic broadcast service when receivers of good engineering design, including adequate selectivity characteristics, are employed. Quiet hours are not usually applicable to cases of telephone, stereo or other home electronic interference, but only to broadcast receiver interference. Even then, there are usually no quiet hours applicable, except when there has been an actual finding that the broadcast receiver is of good engineering design and has adequate selectivity [97.121(a)].

The FCC may impose quiet hours for interference to other radio services, and the ARRL has received sporadic reports of quiet hours imposed by a local field office "in the public interest." These are usually cases where the amateur did not cooperate with the FCC and did not offer any cooperation to bona-fide attempts by a neighbor trying to cure the interference. If the FCC intends to put an amateur on "quiet hours," or to impose other operating restrictions, it is obliged to afford the amateur an opportunity for a hearing before the restrictions are effective [§§303(f) and 316, Communications Act of 1934 as amended].

Living With Your Neighbors

While adhering to the regulations about spurious emissions is your sole legal responsibility, it is usually best to help your neighbors locate solutions for interference problems that involve you and your station. While you may not be the *cause* of fundamental overload EMI problems, you *are* involved with the problem and you may be a necessary part of the solution! If your neighbor contacts the cable company to repair a case of TVI, and you are not available to put your station on the air while the cable repair people try different solutions, the repair staff will probably mark the service form "No Trouble Found" and go on to the next call. You and your neighbor will still have an interference problem. Amateur Radio is a public service, and your technical skills are an important part of that public service. There is no better place to apply those skills than to a local problem that involves you!

Your role, however, is to be a *locator* of solutions, not a *provider* of solutions! Part 97 does not require that you purchase filters for your neighbors' equipment. You may, of course, do so voluntarily, as a neighborly gesture. This may be the easiest way out if you live in a remote area and are causing interference to only a few pieces of equipment. The ham who lives in an apartment complex usually cannot afford to purchase filters for every neighbor with a problem. The manufacturer may be able to help; contact the Electronic Industries Association to locate help at a consumer electronics manufacturer:

Electronic Industries Association
Consumer Electronics Group
Director of Consumer Affairs
2500 Wilson Blvd
Arlington, VA 22201
703-907-7626

You should also help your neighbor understand the instructions that come with the filters used to cure consumer electronics interference problems. You should *not* repair your neighbors' equipment. If you take the back off the TV or work on any equipment, you may be held liable for anything that ever happens to it. (Your neighbor will be happy with the new projection model, of course!)

FCC on RFI in the Neighborhood

An FCC communication supports amateurs on the matter of RFI in the neighborhood and community. The letter, dated November 21, 1991, from

Robert McNamara, then Chief of the FCC Special Services Division, was in response to a complaint from a ham's neighbor:

This is in response to your request for amendment of the Commission's Rules for the Amateur Radio Service. You believe the laws applicable to interference that electronic equipment in a home receives from radio frequency energy are inadequate. Additionally, you disagree with Congress' decision to give the Commission exclusive jurisdiction over interference to home electronic equipment, systems, and devices.

Section 203(a)(2) of the Communications Act of 1934, as amended, 47 USC 203(a)(2), authorizes the Commission to regulate home electronic equipment and systems by establishing minimum performance standards for such equipment to reduce their susceptibility to interference from radio frequency. See 96 Stat. 1087, 1091-1092. The conference Report associated with this section indicates that 'the legislation does not mandate Commission exercise of this authority; that decision is well within the technical expertise of the agency.' The Report also indicates that the Commission, in exercising this authority, is expected to balance the cost of improving the performance of a device against the overall benefit to be gained. See H. Rep. No. 765, 97th Congress, 2d Session (1982), at 32-33. Because most users of home electronic equipment do not receive such interference, we do not wish to impose the additional costs associated with reduced susceptibility on all users of such equipment, including millions of users who would not benefit. Likewise, it is not reasonable to place the burden for resolving all interference problems on amateur service licensees. Congress recognized that electronic equipment manufacturers also have a responsibility to design properly their equipment to prevent interference. We believe the Commission's Rules properly reflect Congressional desires.

The issue of interference to home electronic equipment is being addressed by industry. A committee has been formed under the auspices of the American National Standards Institute to develop, voluntarily, standards to reduce the susceptibility of this equipment to interference. The Commission's long-standing policy, as well as that of the Federal Government in general, is to rely on private industry voluntary standards whenever possible. At our encouragement, the Electronic Industries Association (EIA) developed, in 1984 and 1987, two susceptibility standards for television receivers.

These standards were developed using American National Standards Institute procedures. Recent figures provided by the EIA indicated that virtually all new color televisions and VCRs voluntarily comply with these standards. Additionally, the number of complaints we received about interference to home electronic equipment has dropped sufficiently since 1982. Earlier this year, the Telecommunications Industry Association adopted a standard for telephone terminal equipment that contains product goals and for electromagnetic interference susceptibility. In addition, international standards on interference susceptibility are being developed for a wide variety of electronic products. Although compliance with these standards is voluntary, we expect their development will spur electronic equipment manufacturers to consider potential interference problems when designing their equipment. Interference to the type of electronic equipment you mentioned in your letter does not give the Commission a basis to restrict the operation of your neighbor or modify his license. See Section 15.1, 15.5(b), and 97.121 of the Commission's Rules 47 CFR 15.1, 15.5(b), and 97.121. Additionally, the Communications Act grants a licensee certain rights such as a right to a hearing, before the Commission can modify a station license. See Section 316 of the Communications Act of 1934, as amended, 47 USC, 316. Based on the above, we conclude that your proposal is not in the public interest and does not warrant consideration by the Commission. Accordingly, pursuant to Section 1. 401(e) of the Commission's rules, 47 CFR §1. 401(e), it is ordered that your request for rulemaking is denied.

FCC on Preemption of RFI Matters

Occasionally, a municipality will get interested in regulating RFI in its local ordinances. The Commission recently sent a letter to the ARRL General Counsel's office in response to the ARRL's request for an opinion on a Pierre, South Dakota, ordinance regulating RFI.

The letter from the FCC stated that Congress has preempted any concurrent state or local regulations of RFI pursuant to the provisions of the Communications Act. The act provides that the FCC "may, consistent with the public interest, convenience and necessity, make reasonable regulations (1) governing the RFI potentials of devices which ... are capable of emitting radio frequency energies...in sufficient degree to cause harmful interference to radio communications..." The FCC says the legislation provides explicitly that the Commission has exclusive authority to regulate RFI. Congress declared "the reservation of exclusive jurisdiction to the FCC over matters involving RFI. Such matters shall not be regulated by local or state law, nor shall radio transmitting be subject to local or state regulation as part of any effort to resolve an RFI complaint." The FCC letter said that "state laws that require amateurs to cease operations or incur penalties as a consequence of RFI thus have been entirely preempted by Congress." The Pettit letter is Appendix 8.

Low-Power RF Devices and Interference

On March 30, 1989, the FCC amended Part 15 of its rules governing radio frequency devices, which, because of their low power, are not required to be licensed. There was good, bad and ghastly news for radio amateurs.

The good news is that leakage permitted from some "Unintentional" radiating devices (a broad class of devices including radio and TV receivers, VCRs, stereo equipment and the like) will have to be reduced to the more stringent limits now permitted to Class B computing devices used in the home.

The bad news is that noncompliant devices of this type are grandfathered for 10 years. Existing TV tuners, for instance, can be designed, built and sold until the late 1990s, observing only today's radiation limits. Although the FCC didn't say a word about RFI to consumer devices, the improvements eventually made in front ends of TVs and VCR/TV combinations should result in improved RF rejection, and thus less likelihood of RFI complaints, but we'll have to wait until the next century to know.

The ghastly news is that the FCC also adopted seven new "consumer bands" where "intentional" radiating devices, such as home security systems, garage-door openers, wireless stereo speakers and TV "wireless rabbits" (devices that transfer programs to other TVs and VCRs throughout the house) may operate with higher power than otherwise would be permitted. Four of the specified bands, supposedly selected because Industrial, Scientific and Medical (ISM) devices already operate there, are allocated to the Amateur Radio Service on a primary or secondary basis. These bands are: 902-920 MHz, 2400-2483.5 (the amateur band here is 2390-2450 MHz, and amateurs are primary on the 2390-2400 MHz portion), 5725-5875 MHz (hams have 5650-5925 MHz) and 24.00-24.25 GHz (of which 24.00-24.05 is exclusively amateur, the remainder shared).

Manufacturers are not required to include instructions on dealing with interference in consumer guides furnished with their equipment. The only gesture the FCC made toward warning the public about interference is a label stating that the device must not cause interference to any licensed radio service and must accept interference which occurs to it from such services. The second part of this label is not required on radio and TV receivers.

For amateurs, two important rules apply in matters involving interference to and from "Part 15" devices:

§15.5 General conditions of operation.

(b) Operation of an intentional, unintentional or incidental radiator is subject to the conditions that no harmful interference is caused and that interference must be accepted that may be caused by the operation of an authorized radio station, by another intentional or unintentional radiator, by industrial, scientific and medical (ISM) equipment, or by an incidental radiator.

(c) The operator of a radio frequency device shall be required to cease operating the device upon notification by a Commission representative that the device is causing harmful interference. Operation shall not resume until the condition causing the harmful interference has been corrected.

Cable Television Interference

As with the Amateur Radio Service, the Cable Television Service is regulated by the FCC. Part 76 of the Commission's rules concerns cable system operators. The FCC defines a cable plant as:

"Cable system or cable television system. A facility consisting of a set of closed transmission paths and associated signal generation, reception and control equipment, that is designed to provide cable service which includes video programming and which is provided to multiple subscribers within a community. . ." [76.5(a)].

The operative word here is "closed." Programming is distributed by a system of cables and associated equipment—pathways which, by definition, do not use the airwaves. Because it is supposed to be a closed system, the cable company operator is responsible for cleaning up any interference it generates to your station. It is also responsible for cleaning up "ingress" problems where your signal is allowed to enter the cable system, causing interference to subscribers' TV operation.

Technical Standards for CATV operation

Just as amateurs are required to ensure that their operations meet certain technical standards, cable system operators must also comply with similar FCC-imposed standards. The limits for allowable radiation from a cable system are contained in Section 76.605(a)(11) of the rules:

Frequencies	Radiation Limit ($\mu V/m$)	Distance (meters)
Less than and including 54 MHz	15	30
Over 54 up to and including 216 MHz	20	3
Over 216 MHz	15	30

Interference from a CATVI System

Section 76.613 regulates interference from cable television systems. Paragraph (a) defines harmful interference as "any emission, radiation or induction which endangers the functioning of a radionavigation service or of other safety services or seriously degrades, obstructs or repeatedly interrupts a radiocommunication service operating in accordance with this chapter."

Of critical significance to amateurs experiencing CATVI is paragraph (b):

"(b) The operator of a cable television system that causes harmful interference shall promptly take appropriate measures to eliminate the harmful interference."

Paragraph (c) provides authority to an FCC Engineer-in-Charge (EIC) for the suspension of a cable system operation should harmful interference to radiocommunication involving the safety of life or protection of property not be promptly eliminated by the application of suitable techniques.

Paragraph (d) states that "The cable television system operator may be required by the EIC to prepare and submit a report regarding the cause(s) of the interference, corrective measures planned or taken and the efficacy of the remedial measures."

In addition, the cable company is required to deliver a quality signal to its subscribers. Section 76.605 requires specific minimum signal-to-coherent disturbance ratios for different types of cable systems. If you are leaking *into* the cable, the performance of the system is being degraded and they need to locate and correct the leak.

If You Experience CATVI

If you experience CATVI, the first step is to try to determine the origin of the interfering signals: Find the leak! Then write a letter to the system operator outlining the problem and the steps you have taken thus far, and reminding the company of its obligation under the rules to clean up the interference. Try to seek out someone within the company who has the technical background necessary to deal effectively with the problem. If possible, enlist the support of other amateurs who are experiencing similar interference. Should the cable company adopt an unresponsive or uncooperative attitude, write again outlining the continuing problem, and send a copy of your letter to the local FCC Field Office and to the municipal government or public utility commission exercising local control over the company's operation. It is normally in the best interest of the company to be responsive to complaints, as it can face federally imposed fines and local enforcement action by municipal bodies in franchise agreements and state control in public utility regulation.

You may also seek assistance from the National Cable TV Association (NCTA). The address is:

National Cable TV Association
ATTN: Wendell Bailey, Director of Engineering
Science and Technology Department
1724 Massachusetts Ave, NW
Washington, DC 20036

If you are unsuccessful in finding a solution after the above steps have been taken, contact the FCC Engineer-in-Charge of your local FCC Office, in writing, explaining the steps taken to solve the problem. Ask the FCC to assign a case number for later reference. It is important that the FCC be contacted only as a last resort.

Power-Line Noise Problem

Power-line noise is a problem in some areas. The power company is responsible for any interference it generates to your station or to anyone's receiving equipment in the area. Power line noise comes under the heading of Part 15 devices and the same rules apply as listed earlier in the discussion of low-power RF devices. Inform the general manager of your power company of its obligation under these rules in a registered, return-receipt letter. Keep copies of all correspondence. If you hear nothing after a reasonable period of time (usually 30 days), send a follow-up letter to the general manager of your power company with a copy sent to the president or executive vice president of your power company. If the situation is not corrected to your satisfaction after a reasonable time, send a registered letter to the chairman or commissioner of your state public utilities commission with a copy sent to the Engineer-in-Charge of your local FCC Field Office. The FCC is reluctant to become involved unless all other remedies have been tried.

Protection for FCC Monitoring Facilities

If you live within one mile of an FCC Monitoring Facility (see Appendix 12), you must take steps to ensure you don't harmfully interfere with it. If you do interfere, the FCC can put restrictions on your operating [97.13(b)].

National Radio Quiet Zone

If you're putting up a repeater or automatically controlled beacon in the National Radio Quiet Zone (see coordinates below), you must give written notification of your plan to the Director, National Radio Astronomy Observatory, POB 2, Green Bank, WV 24944. You must also notify the Director if you're planning to modify (change frequency, power, antenna height or directivity) an existing repeater or beacon in the zone.

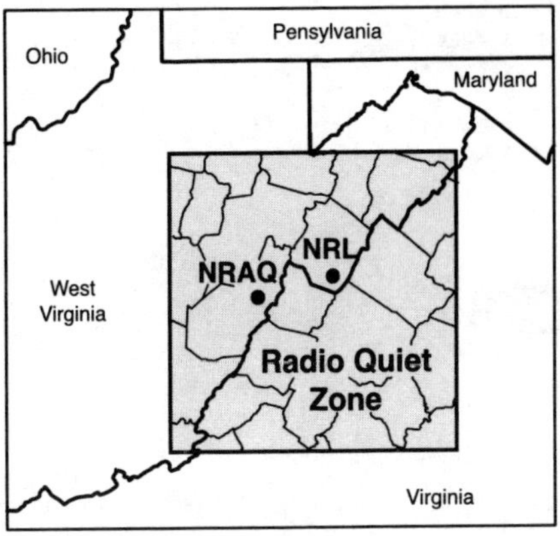

Fig 1—The National Radio Quiet Zone

Notification must include the geographical coordinates of the antenna, antenna height and directivity, proposed frequency, emission type and power. If the NRAO objects within 20 days, the FCC will consider the problem and take appropriate action.

The National Radio Quiet Zone is the area in Maryland, West Virginia and Virginia bounded by 39°15' N, 78°30' W, 37°30' N and 80°30' W. See Fig 1 [97.203(e), 97.205(f)].

Help

The ARRL Technical Information Service has a number of information packages available at nominal cost. These include a general EMI/RFI information package, and more specific ones dealing with CATVI, electrical interference, television interference, telephone interference, interference to audio equipment, and interference from nonamateur transmitters. Send a 9 × 12-inch SASE to the Technical Information Service at ARRL HQ for a list of all currently available packages.

CHAPTER 11

Emissions: Modulation, Signals and Information

Once upon a time, if it sounded like a duck, you had to call it J3E. Thankfully the FCC now just calls it "phone." This approach is used in Part 97 with a cross-reference to the alphanumeric ITU emission designators. For those who are at loose ends unless they know which specific emission they're emitting, Table 1 lists the appropriate designators. The complete list is in §2.201 of the FCC Rules, which can be found in the Appendix.

To find out what is permitted under the emission designators, start by looking up the definitions [97.3(c)]. For CW and test emissions, check 97.305(a) and (b), respectively. For other emission designators, check 97.305(c) to find the bands where these emissions are permitted, along with any restrictions and/or additional modes authorized for specific bands, which are located in 97.307(f). The rules provide latitude for the experimenter, while preserving the traditional demarcation between RTTY/data and phone/FAX/SSTV operation in the HF bands.

Emission Types

The FCC defines nine different types of emissions in §97.3(c): CW, MCW, Phone, Image, RTTY, Data, Pulse, Spread Spectrum (SS) and Test. Differences between these emissions can be confusing at times. The same equipment can be used to transmit Phone, SSTV or RTTY. The type of information transmitted, as well as how it's transmitted, determines how an emission is classified. Let's look at each one.

CW

The term *CW* (Continuous Wave) is used by hams to denote the oldest radio modulation system, telegraphy by on-off keying of a carrier. CW emission can be used on any frequency in the amateur bands [97.305(a)]. The term continuous wave came about to differentiate a carrier generated by sources that produce a clean, steady signal (such as oscillators) from *damped waves*, the rough, broad signals generated by a spark gap transmitter (damped wave transmissions were outlawed in the 1930s). Some parts of the electronics industry use the term CW to signify a steady unmodulated carrier.

In the past, the FCC only used the designator A1A to designate CW (on/off keying of the main carrier), but wouldn't have minded if you produced the same result by keying a tone and feeding it into a single-sideband transmitter. Technically that would have been J2A, which the old rules did not specifically say was permissible. However, the current rules explicitly permit on/off keying of a tone modulating a single-sideband transmitter. As with any CW signal, you need to use a beat-frequency oscillator (BFO) to hear the Morse code keyed tone.

MCW

MCW (Modulated CW) is produced by modulating a carrier with a tone, then keying the tone or both the tone and the carrier. Most ham rigs don't have this mode built in, in which case you supply the tone with an external tone generator. Unlike CW, you don't need a beat-frequency oscillator (BFO) to hear the tone. This makes it particularly useful for repeater IDs.

For many years, MCW was the modulation of choice for emergency transmitters used in aircraft and vessels. MCW occupies a bandwidth wider than that of ordinary CW. With hard keying and square-wave tone modulation, MCW creates such a splatter that it is hard to miss, even with a mistuned receiver, with or without a BFO. MCW, even with soft keying and a clean sine-wave modulating tone, is not permitted in the crowded HF ham bands for general communications, for example as a substitute for CW. There is a special provision in the rules allowing MCW to be used on the HF phone bands for code practice transmissions when interspersed with speech [97.3(c)(5)]. It isn't MCW, however, unless a carrier is transmitted, as in the case of tone keying an FM or AM transmitter. If a keyed tone is transmitted over an SSB transmitter, there is no carrier, so the Morse transmission is CW, not MCW, and a BFO is needed in the receiver.

PHONE

Phone (short for telephony) used to mean speech only. Now it includes speech and other sounds. No, that doesn't mean rock music with 100% distortion, but it does include almost any sound hams might need to transmit. *Amplitude Modulation* (AM, designator A3E), *Frequency Modulation* (FM, designator F3E) and *Single Sideband Suppressed Carrier* (SSB, designator J3E) emissions are all included under phone, as long as sounds are transmitted.

Tone alerting and selective calling are explicitly permitted on frequencies authorized for phone [97.3(c)(5)]. The rules permit subcarrier tones that are incidental to the operation of the telephone channel. For example, a subcarrier to control the level of the demodulated signal (compandoring) is permitted—so you can add a pilot tone for *amplitude-compandored single sideband* (ACSSB) if desired [97.3(c)(5)].

You can also send your station ID by Morse code or other digital means on any frequency where phone operation is permitted. The key word here is *incidental*: Speech is the main thing, and the tones are supplemental [97.3(c)(5)].

You can choose virtually any phone modulation method that can convey speech: double-sideband full carrier, single-sideband full carrier, single-sideband reduced or variable-level carrier, single-sideband suppressed carrier, vestigial sideband, or independent sidebands with one sideband carrying speech and the other having slow-scan TV or facsimile. The last is a form of multiplexing—carrying two or more channels of information on one signal.

You are no longer restricted to analog modulation—digitized signals are also permitted, as long as the intelligence being transmitted is telephony. Thus, digital voice is permitted wherever phone operation is authorized.

Is there any bandwidth limitation to phone transmissions? None according to the rules, except for the general requirement of "good engineering practice" [97.101(a)]. On the other hand, using excessive bandwidth will get you in immediate trouble with other hams who are trying to share the band with you. So it's wise to limit phone bandwidth on the HF bands to that of a single-sideband transmission of a single analog audio channel (roughly 3 kHz) if possible. There are exceptions to that: bandwidths on the order of 6 kHz are needed for double-sideband AM phone, for FM phone and for independent-sideband transmissions, where each sideband is roughly 3 kHz wide. Good engineering practice in these cases boils down to "reasonable and proper bandwidth"—don't use more than you need.

Above 902 MHz, the rules permit multiplexing (MUXing) FM phone modulation (designated F8E) [97.305(c); 97.307(f)(12)]. This permits more than one analog channel to be transmitted on the main carrier. Good engineering practice for multiplexing again amounts to using no more bandwidth than you need. When in doubt, check with the frequency coordinator in your area to see what sort of protection adjacent-channel users need.

IMAGE

The word *image* encompasses both television and facsimile. Basically, the rules permit you to transmit image information using essentially any modulation method. Because image operation on HF is within the phone bands, you can alternate image modulation with phone modulation. You can also transmit image and phone on the same carrier by using independent-sideband modulation. The bandwidth limitation in this case is essentially the same as that for transmitting a single channel of communications quality speech, or 3 kHz for single sideband, 6 kHz for double-sideband or for two channels of information [97.305(c), 97.307(f)(2)]. If you can fit it into this bandwidth, you can transmit it. Both *slow scan television* (SSTV) and *facsimile* (FAX) can be transmitted here.

The rule limiting bandwidth of an image emission to that of a communications quality phone signal of the same modulation type applies to all amateur bands from 160 to 1.25 meters. Thus fast scan television, with its wide bandwidth requirement, is permitted only in and above the 70 cm (440 MHz) band.

Image transmissions may be either analog or digital, as long as they meet the bandwidth limitations for the frequencies to be used. Station ID may be done by image as long as all or part of the communications on the frequency are in the same image emission, and it conforms to the applicable standards of §73.682(a) (NTSC transmissions).

RTTY AND DATA

What's the difference between *RTTY* and *data*? There may be some gray areas, but here are some characteristics to help you differentiate between the two:

RTTY

   Keyboard sending, hard copy or screen presentation, possibly with computer buffering
   Little or no manipulation or reformatting of text prior to presentation (what is sent is what you get)
   Text in Baudot or ASCII
   Has no error control (except when using AMTOR)
   Message transmission

Asynchronous transmission (in other words, start and stop pulses are used, except when using AMTOR)

Data

Computer file sending, computer file receiving

Possibly some reorganization of information prior to transmission and probably before presentation

If text, probably in ASCII, Extended ASCII, or foreign language character set

Usually has error detection and/or error correction

Probably packetized transmission

Synchronous transmission (no start and stop pulses)

It probably makes little or no difference whether there's a gray area between RTTY and data transmission, however, since the rules permit both in the same frequency segments. What types of data transmission are permissible under the rules? First, you can transmit specified digital codes of Baudot, AMTOR and ASCII [97.309(a)]. A station may also use an "unspecified digital code, except to a station in a country with which the United States does not have an agreement permitting the code to be used" [97.309(b)]. "Unspecified digital codes must not be transmitted for the purpose of obscuring the meaning of any communication" [97.309(b)]. The rules permit unspecified digital codes as well as the previously mentioned "specified" codes to be used above 50.1 MHz, but only Baudot, AMTOR or ASCII may be used below that frequency [97.305(c) and 97.307(f)(3),(4)]. A discussion of digital codes is found in Chapter 13. RTTY and data multiplexing is allowed on frequencies above 50.1 MHz within the bandwidth limitations [97.307(f)(5)-(8)].

PULSE

Pulse is actually a type of modulation of the main carrier—on a par with amplitude modulation (AM) and angle modulation (a category covering both frequency modulation and phase modulation) [97.3(c)(6)]. Pulses may be modulated in the following ways:

• unmodulated pulses
• pulses modulated in amplitude
• pulses modulated in width/duration
• pulses modulated in position/phase
• pulses in which the carrier is angle modulated during the period of the pulse

Pulse emission also permits transmission of any type of information, namely:
• no information
• telegraphy
• facsimile
• data transmission, telemetry, telecommand
• telephony
• television
• combination of the above (3rd symbol W)
• cases not otherwise covered (3rd symbol X)

Pulse is permitted in bands above 902 MHz excluding 1240-1300 MHz and 10.0-10.5 GHz [97.305(c)].

SS

Spread spectrum (SS) is a departure from the normal rule of keeping radio-frequency bandwidth as narrow as possible for a given information rate. SS deliberately expands the bandwidth; instead of concentrating all the transmitted energy in a tight bandwidth, the energy is spread over a greater bandwidth. The bandwidth expansion amounts to at least 10 times the information bandwidth and may be as much as several hundred times.

Checking the ITU designators shows that you can use any type of modulation of the carrier (except for no modulation or pulse modulation) and can transmit any type of signal and information permitted in the Amateur Radio Service when using SS.

There are some restrictions. First, SS may be used only on bands above 420 MHz [97.305(c)]. Code, modulation and transmitter power limitations are detailed in 97.311. See also "Spread Spectrum" in Chapter 12.

TEST

When the FCC rewrote Part 97 in 1989, they tried to clear up past uncertainties concerning emission of experimental transmissions and those with no information. "No information" is not necessarily the same as "no modulation," as it is possible to frequency modulate a transmitter with a steady tone that carries no intelligence. We're talking about transmissions like unmodulated tones or key-down carriers to tune transmitters or to measure antenna patterns, for example. One of the most common uses of test transmissions above 51 MHz is for alignment of high gain antennas.

The FCC decided to use the emission designator test as a catchall category. Technically, test is defined as "Emissions containing no information having the designators with N as the third symbol" [97.3(c)(9)].

There are two basic rules of the road for test emissions. Test emissions are authorized for all amateur bands above 51.0 MHz [97.305(c)], and there is no limit to the length of your test signals. It's uncertain how other amateurs would welcome your transmitter being key down 24 hours a day, however. It may take quite a long time to align antennas on a 10-GHz circuit, but it's not a good idea to do so on your local repeater. Below 51.0 MHz, however, §97.305(b) applies, and the key word is *brief*.

Table 1
ITU Designations

CW

The FCC uses these ITU designators for CW: International Morse code telegraphy emissions having designators with A, C, H, J or R as the first symbol; 1 as the second symbol; A or B as the third symbol; J2A or J2B [97.3(c)(1)].

First symbol	Second symbol	Third symbol
Type of modulation of the main carrier:	*Nature of signal(s) modulating the main carrier:*	*Type of information to be transmitted:*
A Double sideband **C** Vestigial sideband **H** Single sideband, full carrier **J** Single sideband, suppressed carrier **R** Single sideband, reduced carrier	**1** A single channel containing quantized or digital information without the use of a modulating subcarrier	**A** Telegraphy—for aural reception **B** Telegraphy—for automatic reception

And J2A and J2B:

First symbol	Second symbol	Third symbol
J Single sideband, suppressed carrier	**2** A single channel containing quantized or digital information with the use of a modulating subcarrier	**A** Telegraphy—for aural reception **B** Telegraphy—for automatic reception

MCW

This is what the rules permit as MCW emission: Tone-modulated international Morse code telegraphy emissions having designators with A, C, D, F, G or R as the first symbol; 2 as the second symbol; A or B as the third symbol [97.3(c)(4)].

First symbol	Second symbol	Third symbol
Type of modulation of the main carrier:	*Nature of signal(s) modulating the main carrier:*	*Type of information to be transmitted:*
A Double sideband **C** Vestigial sideband **D** Emission in which the main carrier is amplitude and angle modulated either simultaneously or in a pre-established pattern **F** Frequency modulation **G** Phase modulation **R** Single sideband, reduced carrier	**2** A single channel containing quantized or digital information with the use of a modulating subcarrier	**A** Telegraphy—for aural reception **B** Telegraphy—for automatic reception

Phone

The following is the ITU definition of Phone emission: Speech and other sound emissions having designators with A, C, D, F, G, H, J or R as the first symbol; 1, 2 or 3 as the second symbol; E as the third symbol. Also speech emissions having B as the first symbol; 7, 8 or 9 as the second symbol; E as the third symbol [97.3(c)(5)].

First symbol

Type of modulation of the main carrier:

A Double sideband
C Vestigial sideband
D Emission in which the main carrier is amplitude and angle modulated either simultaneously or in a pre-established pattern
F Frequency modulation
G Phase modulation
H Single sideband, full carrier
J Single sideband, suppressed carrier
R Single sideband, reduced carrier

Second symbol

Nature of signal(s) modulating the main carrier:

1 A single channel containing quantized or digital information without the use of a modulating subcarrier
2 A single channel containing quantized or digital information with the use of a modulating subcarrier
3 A single channel containing analog information

Third symbol

Type of information to be transmitted:

E Telephony

Also:

B Independent sidebands

7 Two or more channels containing quantized or digital information
8 Two or more channels containing analog information
9 Composite system with one or more channels containing quantized or digital information, together with one or more channels containing analog information

E Telephony

And F8E above 902 MHz:

F Frequency modulation 8 Two or more channels containing analog information E Telephony

Image

The FCC uses these ITU designators for Image: Facsimile and television emissions having designators with A, C, D, F, G, H, J or R as the first symbol; 1, 2 or 3 as the second symbol; C or F as the third symbol; and emissions having B as the first symbol; 7, 8 or 9 as the second symbol; W as the third symbol [97.3(c)(3)].

First symbol

Type of modulation of the main carrier:

A Double sideband
C Vestigial sideband
D Emission in which the main carrier is amplitude and angle modulated either simultaneously or in a pre-established pattern
F Frequency modulation
G Phase modulation
H Single sideband, full carrier
J Single sideband, suppressed carrier
R Single sideband, reduced carrier

Also:

B Independent sidebands

Second symbol

Nature of signal(s) modulating the main carrier:

1 A single channel containing quantized or digital information without the use of a modulating subcarrier
2 A single channel containing quantized or digital information with the use of a modulating subcarrier
3 A single channel containing analog information

7 Two or more channels containing quantized or digital information
8 Two or more channels containing analog information
9 Composite system with one or more channels containing quantized or digital information, together with one or more channels containing analog information

Third symbol

Type of information to be transmitted:

C Facsimile
F Television (video)

W Combination of the above

"W Combination of the Above" covers a combination of facsimile or television and telephony by means of independent-sideband (ISB) modulation. ISB modulation with image on one sideband and phone on another is permitted in the MF and HF bands. Literally, the designator means a combination of any of the following types of information:

    N No information transmitted

A Telegraphy—for aural reception
B Telegraphy—for automatic reception
C Facsimile
D Data transmission, telemetry, telecommand
E Telephony
F Video

However, when transmitting ISB, it would be prudent to limit the information content of the two independent sidebands to the types of information permitted in that band segment. As phone and image are usually permitted in the same segments and CW is allowed in all bands, it would seem that these are the only emissions permitted for ISB modulation in such segments.

RTTY & Data

RTTY

   Here's what the rules permit under the RTTY emission: Narrow-band direct-printing telegraphy emissions having designators with A, C, D, F, G, H, J or R as the first symbol; 1 as the second symbol; B as the third symbol; and emission J2B [97.3(c)(7)]. In terms of ITU emission designator definitions, that means:

First symbol	**Second symbol**	**Third symbol**
Type of modulation of the main carrier:	*Nature of signal(s) modulating the main carrier:*	*Type of information to be transmitted:*
A Double sideband	**1** A single channel containing quantized or digital information without the use of a modulating subcarrier	**B** Telegraphy—for automatic reception
C Vestigial sideband		
D Emission in which the main carrier is amplitude and angle modulated either simultaneously or in a pre-established pattern		
F Frequency modulation		
G Phase modulation		
H Single sideband, full carrier		
J Single sideband, suppressed carrier		
R Single sideband, reduced carrier		

and J2B:

J Single sideband, suppressed carrier	**2** A single channel containing quantized or digital information with the use of a modulating subcarrier	**B** Telegraphy—for automatic reception

Data

In terms of ITU emission designators, the following types of emission are permitted for data emission: Telemetry, telecommand and computer communications emissions having designators with A, C, D, F, G, H, J or R as the first symbol; 1 as the second symbol; D as the third symbol; and emission J2D. That expands to:

First symbol
Type of modulation of the main carrier:

A Double sideband
C Vestigial sideband
D Emission in which the main carrier is amplitude and angle modulated either simultaneously or in a pre-established pattern
F Frequency modulation
G Phase modulation
H Single sideband, full carrier
J Single sideband, suppressed carrier
R Single sideband, reduced carrier

Second symbol
Nature of signal(s) modulating the main carrier:

1 A single channel containing quantized or digital information without the use of a modulating subcarrier

Third symbol
Type of information to be transmitted:

D Data transmission, telemetry, telecommand

and J2D:

J Single sideband, suppressed carrier

2 A single channel containing quantized or digital information with the use of a modulating subcarrier

D Data transmission, telemetry, telecommand

Spread Spectrum

The rules [97.3(c)(8)] permit the following types of modulation, according to ITU emission designators: A,C,D,F,G,H,J or R as the first symbol, X as the second symbol, and X as the third symbol.

First symbol

Type of modulation of the main carrier:

A Double sideband
C Vestigial sideband
D Emission in which the main carrier is amplitude and angle modulated either simultaneously or in a pre-established pattern
F Frequency modulation
G Phase modulation
H Single sideband, full carrier
J Single sideband, suppressed carrier
R Single sideband, reduced carrier

Second symbol

Nature of signal(s) modulating the main carrier:

X Cases not otherwise covered

Third symbol

Type of information to be transmitted:

X Cases not otherwise covered

CHAPTER 12
Technical Standards

L ike most of Part 97, the technical standards exist to promote operating techniques that make efficient use of spectrum and minimize interference. No one wants to transmit a "dirty" signal, and the standards in Part 97 can help you identify problems that should be solved.

Good Signals: 99-44/100% Pure

Section 97.307 spells out the standards the FCC expects amateur transmissions to meet. The first paragraph is the most important: "No amateur station transmission shall occupy more bandwidth than necessary for the information rate and emission type being transmitted, in accordance with good amateur practice" [97.307(a)]. Simply stated, don't transmit a wide signal when a narrow one will do.

The next paragraph states that "Emissions resulting from modulation must be confined to the band or segment available to the control operator." Every modulated signal produces sidebands. You must not operate so close to the band edge that your sidebands extend out of the subband, even if your dial says that your carrier is inside the band. Further: "Emissions outside the necessary bandwidth must not cause splatter or key-click interference to operations on adjacent frequencies" [97.307(b)]. Seems like common sense, doesn't it? The rules simply codify good operating practice. If your key clicks or over-processed voice signals are causing interference up and down the band, clean them up! You wouldn't ask any less of another ham with a messy signal.

Spurious Emissions

The rules address *spurious emission* standards [97.307(c)]. Just what are spurious emissions anyway? The FCC defines them this way: "An emission, on frequencies outside the necessary bandwidth of a transmission, the level of which may be reduced without affecting the information being transmitted" [97.3(a)(40)]. In addition, Part 2 of the FCC Rules adds the following (from the international Radio Regulations): "Spurious emissions include harmonic emissions, parasitic emissions, intermodulation products and frequency conversion products, but exclude out-of-band emissions" [2.1(c)].

To fully understand this, we must again consult Part 2 for more definitions. *Necessary bandwidth* is defined as: "the width of the frequency band which is just sufficient to ensure the transmission of information at the rate and with the quality required under specified conditions." An *out-of-band emission* is: "Emission on a

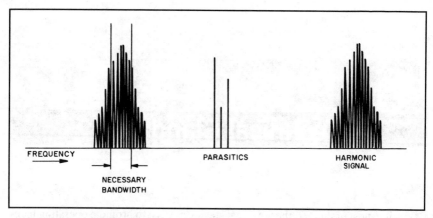

Fig 1—An illustration of spurious emissions. Some of the keying sidebands are outside the necessary bandwidth. These are considered out-of-band emissions, not spurs. The harmonic and parasitic shown here are considered spurioius emmisions; these must be reduced in accordance with §97.307.

frequency or frequencies immediately outside the necessary bandwidth which results from the modulation process, but excluding spurious emissions." Out-of-band emissions are not necessarily outside an amateur "band" [2.1(c)].

Refer to Fig 1. Some of the keying sidebands are outside the "necessary bandwidth." These are out-of-band emissions, but they are not considered spurious emissions. On the other hand, we have already seen [97.307(b)] that these sidebands must not interfere with other stations. The harmonics and parasitics shown in Fig 1 are spurs, and they must be reduced to the levels specified in Part 97. The FCC states that all spurious emissions must be reduced "to the greatest extent practicable." Further, "if any spurious emission, including chassis or power line radiation, causes harmful interference to the reception of another radio station, the licensee of the interfering amateur station is required to take steps to eliminate the interference." If your spurs are causing interference, it's your job to clean them up [97.307(c)]. Make sure your transmitter is clean!

How Far is Far Enough?

Now that we know what the spurs are, what do we do with them? The FCC is very specific [97.307(d)]. If your transmitter or RF power amplifier was built after April 14, 1977, or first marketed after December 31, 1977, and transmits on frequencies below 30 MHz the mean power of any spurious emissions must:

• never be more than 50 mW;
• be at least 30 dB below the mean power of the fundamental emission, if the mean power output is less than 5 W; and
• be at least 40 dB below the mean power of the fundamental emission, if the mean power output is 5 W or more. See Fig 2.

The following requirements apply between 30 and 225 MHz [97.307(e)]:

• Transmitters with 25 W or less mean output power: spurs must be at least 40 dB below

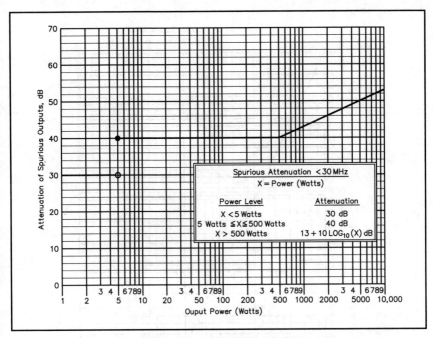

Fig 2—Required attenuation of spurious outputs below 30 MHz is related to output power.

the mean power of the fundamental emission and never greater than 25 μW, but need not be reduced further than 10 μW. This means that the spurs from a 25 W transmitter must be at least 60 dB down to meet the 25 μW restriction.

• Transmitters with more than 25 W mean output power: spurious emissions must be at least 60 dB below the mean power of fundamental emission. See Fig 3.

Transmitter Power

"An amateur station must use the minimum transmitter power necessary to carry out the desired communication" [97.313(a)].

In other words, don't run more power than you need.

How closely do radio amateurs adhere to this principle? We all know that in some cases the answer is, not as closely as we should.

We know there are times when we could get along with reduced power. Even for moonbounce you don't always need to run the legal limit. Most hams don't own legal limit equipment and have no interest in doing so. Many enjoy the challenge of QRP operating, and some can boast of making intercontinental contacts at milliwatt levels. Others must run low power for economic or other reasons, but still gain a full measure of enjoyment from Amateur Radio.

When you reduce power, you reduce the impact your operating has on others. Nearby amateurs can use the same band; distant amateurs can use the same or adjacent frequencies. Your electric bill goes down. If the person you're talking to was receiving you well to begin with, all that happens when you reduce power is that his S meter doesn't bounce quite as high.

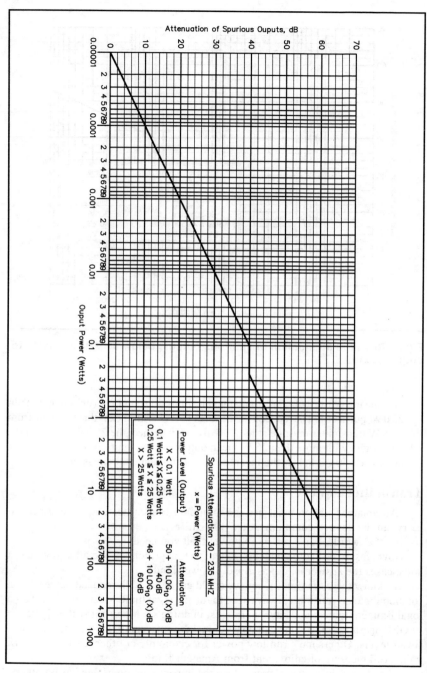

Fig 3—Required attenuation of spurious outputs, 30-235 MHz.

PEP and Power Measurement

Peak envelope power (PEP) is the "average power supplied to the antenna transmission line by a transmitter during one radio frequency cycle at the crest of the modulation envelope taken under normal operating conditions" [2.1(c)]. Amateur transmitters may never be operated with a peak envelope power (PEP) output of more than 1500 W [97.313(b)]. There are further limitations on operation in certain bands, or by holders of certain classes of license. These are detailed below and in Chapter 5.

The FCC has published the following standards of measurement: (1) Read an in-line peak-reading RF wattmeter that is properly matched to the transmission line (commercial units are available), or (2) calculate the power using the peak RF voltage as indicated by an oscilloscope or other peak-reading device. Multiply the peak RF voltage by 0.707, square the result and divide by the load resistance. The SWR must be 1:1. The FCC requires that you meet the power output regulations, but does not require that you make such measurements or possess measurement equipment. The methods listed simply indicate how the Commission would measure your transmitter's output during a station inspection.

As a practical matter, most hams don't have to worry about special equipment to check their transmitter's output because they never approach the 1500-W PEP output limit. Many common amateur amplifiers aren't capable of generating this much power. However, if you do have a planet-destroyer amplifier and do operate close to the limit (only on those rare occasions when you need the extra power, of course) you should be prepared to measure your output along the lines detailed above.

AM Power

When the current power rules went into effect, double-sideband/full-carrier AM users were allowed to use the old measure of 1000 W dc input to the transmitter's final amplifier until June 2, 1990. The reason for this "grandfathering" was that the 1500 W PEP output limit meant an approximate 3-dB reduction in maximum power permitted for AM operations. The FCC has declined to extend this "grandfathering," and AM operators must observe the new standard.

Novice/Technician Plus Subbands

All licensees operating in the Novice/Technician Plus subbands on 80, 40 and 15 meters must never operate with a peak envelope power output in excess of 200 W [97.313(c)(1)]. In addition, Novice and Technician Plus licensees are limited to 200 W PEP output in their 28-MHz subband (28.1-28.5 MHz) [97.313(c)(2)]; Novices are limited to 25 W PEP in the 222 MHz band [97.313(d)] and to 5 W PEP in the 1270 MHz band [97.313(e)]. Other amateurs may use up to 1500 W PEP in these subbands [97.313(c)(2)]. All amateurs are limited to 200 W PEP when operating on the 7.050-7.075 MHz segment in ITU Regions 1 and 3 [97.313(c)(3)].

Special Restrictions

- All amateurs are limited to 200 W PEP on the 30-meter band (10.1-10.15 MHz) [97.313(c)(1)].
- Beacon stations are limited to 100 W power output [97.203(c)].
- 70 cm and 33 cm bands: Special 50 W PEP output limits apply to stations operating in these bands when they are located near certain military installations in the US [97.313(f),(g)].

Q&A—Band Edges

"How close can I set my dial to the band edge?" is a question asked not only by the beginner, but by experienced amateurs as well. We'll tell you how close you can go to get that rare DX QSL card—and avoid that rare FCC "QSL" card.

Phone

Q. I'm an active HF phone DXer. How close to the band edge can I set my VFO?

A. Many factors are involved. Amateurs commonly consider full-carrier, double-sideband AM signals to be about 6 kHz wide and single-sideband, suppressed-carrier signals to be about 3 kHz wide. Those bandwidths, however, are usually only 6 dB down, and that isn't what the FCC worries about. Thus, to determine where you may set your VFO in relation to the band or subband edge for your class of license, you'll have to figure out where your signal is attenuated by 40 dB. Don't assume that if your SSB transmitter bandwidth is 3 kHz, that you can just add a few hundred Hz to be safe. That's fine only if the shape of the band-limiting filter is sharp enough to attenuate the signal to −40 dB at 3 kHz. If you're going to operate as near as possible to a band edge, then do so only after carefully reviewing your equipment specifications to see if such information is included. Another alternative is to carefully measure the attenuation if you have access to top-notch lab equipment. If you can't do either, then allow a larger margin of safety for possible inaccuracy of your frequency readout.

Remember, your carrier and all of your sidebands must be confined within the amateur bands and subbands as applicable.

- A station being used to control a model craft is limited to 1 (yes, that's ONE) W power output [97.215(c)].
- Spread Spectrum stations may not exceed 100 W power output [97.311(g)].

Equipment Type Acceptance

Like other radio services, amateur equipment must meet certain standards. Unlike other services, however, most amateur equipment does not have to be *type accepted*. Type acceptance is an equipment authorization granted by the FCC, based on equipment-measurement data submitted by the applicant, usually the manufacturer. It's used to ensure that the equipment will function properly in the service for which it's been accepted. Type acceptance is usually required for equipment operating in commercial services, such as the Maritime, Public Safety, or Land Mobile services. Even though amateur equipment can often be modified to operate in these services, it is not type accepted and thus may not be used in them.

Morse Code

Q. When I operate CW, how close to the band edge can I go?

A. Your transmitter keying envelope rise and fall times determine both the bandwidth of your CW signal and the maximum speed at which you can send and still be received properly. Once set, the rise and fall times (not the keying speed) determine the bandwidth. If your rise and fall times are 5 ms, your signal will have a bandwidth of about 150 Hz, in which case you would have to stay at least 75 Hz from a band edge. If the rise and fall times are 2 ms (fairly typical of commercial transceivers), the bandwidth would be 375 Hz.

TV and Facsimile

Q. Why is fast-scan television not found on the HF bands?

A. Fast-scan television, a mode in which images appear in the same manner as a home broadcast TV, is not found at HF because of bandwidth limitations. Section 97.307(f)(2) states that on any band below 420 MHz "No non-phone emission shall exceed the bandwidth of a communications quality phone emission of the same modulation type" [97.307(f)(2)]. To get across all the information necessary for a fast-moving TV picture, a great deal of spectrum is required; so much, in fact, that one fast-scan TV signal could occupy one-fifth of the entire HF spectrum! Bandwidth limits restrict HF TV operation to the slow-scan variety, where images appear as a photograph, or "stop action," and do not require oodles of band space.

No special bandwidth limits apply above 420 MHz, so you'll find wider bandwidth modes, such as fast-scan television, there. But remember, in all cases not specifically covered by the rules, the various signals must be used in accordance with good engineering practice. Use modern equipment that is properly adjusted. Conserve spectrum!

Standards for type acceptance are usually higher than those for the amateur service; ham equipment that easily meets amateur standards often cannot meet the more stringent type acceptance requirements. It is okay, however, to modify amateur equipment for use on Military Affiliate Radio System (MARS) or Civil Air Patrol (CAP) frequencies, as long you have authorization to operate there.

Using Type Accepted Equipment on Amateur Frequencies

Equipment which is type accepted for use in another service may be used on amateur frequencies by a licensed amateur as long as it meets all appropriate standards. Modification of type accepted equipment for use on amateur frequencies is likely to void its type acceptance, however, which means that it cannot then be used back in its original service.

Amateur Equipment That Requires Type Acceptance

There is one type of amateur equipment that does require type acceptance. In

1978, the Federal Communications Commission banned the manufacture and marketing of any external radio-frequency power amplifier or amplifier kit that is capable of operation on any frequency below 144 MHz *unless* the FCC has issued a grant of type acceptance for that model amplifier. This was done to stem the flow of amplifiers being distributed for illegal use on frequencies in and around the Citizen's Band Radio Service. The requirements for type acceptance of a commercial amplifier specify that it cannot be able to operate between 24 and 35 MHz, nor can it have accessible wiring, circuitry, internal or external controls, or instructions that will allow it to be operated in a manner contrary to the FCC Rules. It must also meet the emission standards of Part 97 [97.317].

An external RF power amplifier does not need to be type accepted when

- The device isn't capable of operation below 144 MHz.
- The amplifier was purchased before April 28, 1978.
- The amplifier was constructed (but not from a kit) or modified by an amateur for use at that amateur's own station.
- The amplifier is sold to another amateur or to a dealer.
- The amplifier is purchased in used condition by an equipment dealer from an amateur operator and the amplifier is further sold to another amateur operator for use at that operator's station.

An amateur may not construct or modify more than one unit of the same model amplifier capable of operation below 144 MHz in any calendar year without a grant of type acceptance [97.315(a)].

CHAPTER 13

Digital Communications: RTTY and Data

Digital communications are an entire class of emissions that use ones and zeros or discrete (quantized) levels, rather than continuously variable quantities as in analog communications. Although digital communications are associated with computers and machine information transfers, there will be a tendency for some of the older analog modulation techniques to give way to digital techniques for reasons of noise and error reduction. The rules permit a new class of digital emissions for speech and image transmissions. We will not discuss them further in this chapter because we need to wait and see how digital speech and image transmissions develop among amateurs. As these new modes evolve, it will be increasingly difficult to tell them apart from packet radio used to transfer computer data. Computers can, after all, be used to generate speech and images.

RTTY

RTTY is an abbreviation for radioteletype. In §97.3(c)(7) the FCC defines RTTY as "narrow-band direct-printing telegraphy." Direct printing means that the originating station formats the message to print out on a page of paper (hard copy). The receiving operator could elect to display the incoming message on a computer screen. Or, the receiving station computer could store the information in memory and never print it out. Nevertheless, the original intent was to produce page copy at the receiving end.

The bandwidth of a RTTY signal is normally narrower than that of an SSB signal on the MF and HF bands (160-10 meters). The necessary bandwidth for a typical 45-baud (*baud* is a unit of signaling speed equal to one pulse per second) RTTY signal using frequency-shift keying with a shift of 170 Hz is 249 Hz. For information on calculating bandwidth, see the *ARRL Handbook for Radio Amateurs* or the ARRL's *Your HF Digital Companion*.

Data

The rules define *data* as "Telemetry, telecommand and computer communications emissions" [97.3(c)(2)]. That agrees closely with the ITU definition, which is: "Data transmission, telemetry, telecommand" [2.201(e)]. The term "computer communications" is simply another way of saying "data transmission." You are using data transmissions when you exchange computer files, programs and messages not intended for direct paper printout.

Telemetry refers to one-way transmissions of measurements at a distance from the measuring instrument [97.3(a)(43)]. The best-known kind is space telemetry, which is a one-way transmission from a space station (an OSCAR satellite, for example) made from the measuring instruments aboard a spacecraft, including those related to its functioning. An increasing use is gathering weather data. Hams have been remotely controlling repeaters for years with an indication back from the repeater station that it understood the command. After all, if you're going to control something remotely, you better know about its health and welfare before you command it. Once you command it, you'd like to know whether it did what you told it to do. That feedback channel is telemetry. Telemetry is also just measuring anything at a distance, even when you can't remotely control the function. An example of this might be an indication of voltage on a battery charged by solar cells.

Telecommand is the use of telecommunication for the transmission of signals to initiate, modify or terminate functions of equipment at a distance (in other words: remote control) [97.3(a)(41)]. One way to send telecommand or remote-control signals without getting into the realm of unspecified codes would be to use the control characters in ASCII or those positions in Baudot/AMTOR codes not assigned to letters, numerals or fraction bar (/).

Digital Codes

In Part 97, the FCC specifies the following codes for RTTY or data emissions.

Baudot

Baudot is the familiar name for a 5-bit code known worldwide as International Telegraph Alphabet No. 2 (ITA2) [97.309(a)(1)]. Its formal definition is found in International Telegraph and Telephone Consultative Committee (CCITT) Recommendation F.1, Division C. Combinations of ones and zeros in 5 bits can produce 32 positions. Each position is assigned a graphic character or a machine control function. By having two cases (letters and figures), the letters of the Latin alphabet, numerals and commonly used punctuation marks can be accommodated. For a table showing the bit patterns and their assignments, see the *ARRL Handbook*.

AMTOR

AMTOR is an acronym for *AMateur Teleprinting Over Radio*. It is an adaptation of *SITOR (Simplex Teleprinting Over Radio)*, a teletype mode used in the maritime service. The term *TOR* has caught on as a generic reference to these kinds of systems. Its formal definition may be found in International Radio Consultative Committee (CCIR) Recommendations 476-2 (1978), 476-3 (1982), 476-4 (1986) or 625 (1986) [97.309(a)(2)]. The bit patterns and their assignments are given in the *ARRL Handbook*. One important ingredient in TOR is its 7-bit code. For example, the bit sequences for the letter A (and hyphen in figures case) is 1000111, and for B (? in figures case) is 1110010. Both bit sequences have four 1s and three 0s; this is true for all TOR code combinations. This is done to provide a method of error detection at the receiving station. If anything but the correct four 1s-three 0s combination is received, the receiving station knows there was an error in reception.

Another feature is a mode called Automatic Repeat reQuest (ARQ), also called Mode A. This is a protocol that calls for the sending station to transmit three characters, wait for an ACKnowledgment, then three more characters, and so forth. If the four 1s-three 0s combination is not received, an ACK is not given, and the sending station repeats the three characters until an ACK is received.

There is also a Mode B, which uses Forward Error Correction (FEC). Mode B is used for transmissions from one point to more than one receiving station—situations where it would be impractical to wait for an ACK from each receiving station. In the FEC mode, each block of three characters is sent twice in a one-way transmission. The theory is that the receiving station equipment will accept characters having the correct 4/3 ratio and reject the others, and each character gets two chances to be received. Mode B is useful for bulletin transmissions and roundtables where ACKs are not feasible. For example, W1AW transmits its AMTOR bulletins using Mode B.

ASCII

Officially, *ASCII* is called the American Standard Code for Information Interchange by the American National Standards Institute (ANSI) X3.4-1977 [97.309(a)(3)]. The rules also permit International Telegraph and Telephone Consultative Committee (CCITT) Recommendation T.50, International Alphabet No. 5 (IA5), or International Organization for Standardization, International Standard ISO 646 (1983), which are international forms of ASCII [97.309(a)(3)].

ASCII is a 7-unit coded character set having 128 code combinations. A table showing the character set is given in the *ARRL Handbook*.

The rules permit ASCII code extensions as provided for in CCITT Recommendation T.61 (Malaga-Torremolinos, 1984) [97.309(a)]. These extensions provide for additional graphic symbols and diacritical marks used for foreign languages. None of these special characters would come up in everyday English-language text, but they're there if you need them. Permission to use code extensions does not relieve a station of identifying properly using the basic character set. That's the same idea as for phone, in which you can chat in a foreign language, but you must identify in English or international phonetics so others on the band can understand.

Unspecified Digital Codes

Above 50 MHz where data and RTTY emissions are permitted, unspecified digital codes may be used in communication between stations licensed by the FCC [97.307(f);97.309(b)]. Baudot, AMTOR and ASCII are specified in the rules, so unspecified includes everything else. There is a provision for using unspecified codes when communicating with stations in other countries that have an agreement with the US to that effect. However, there are no such agreements in effect as of this writing. Unspecified digital codes must not be used to obscure the meaning of any communication [97.309(b)].

Packet Radio

Packet radio is simply a packaging method for data transmission using the ASCII digital code. The basic idea is to break up a message into small transmission blocks, called packets. Depending on the speed of transmission, a packet may last only a fraction of a second or several seconds. The length may vary, but a typical packet might contain a line of text, a header showing who it came from and where it's going, and a trailer with error-control bits.

Packet radio is one of the most popular modes of digital communications because it is very efficient and provides error-free data transfer. Packet networks can be created using one of several different protocols, allowing long distance communications with limited resources. The *ARRL Handbook* gives a technical description of packet radio and *The ARRL Operating Manual* contains operating information. The ARRL publication *Your Packet Companion* is also very helpful.

Automatic Control of Data and RTTY Stations

An amateur station may be under automatic control when transmitting digital communications on amateur bands 6 meters and above [97.221(b)]. A station may not be operated under automatic control while transmitting third-party communications, except for a station transmitting a RTTY or data emission [97.109(e)].

In 1986 the ARRL requested that the FCC grant a special temporary authority (STA) to allow experimentation with automatic control of certain stations engaged in automatic forwarding of traffic on the HF bands. This operation, known as SKIPNET, served as a test bed for experiments with message-forwarding protocols and station-control techniques. As a result of these studies, the League petitioned the FCC to allow automatic control of digital stations in narrow segments of the HF bands.

On July 1, 1995, the FCC modified its rules to allow automatically controlled HF stations transmitting RTTY and data to communicate with one another in the following segments: 3.620-3.635 MHz; 7.100-7.105 MHz; 10.140-10.150 MHz; 14.095-14.0995 MHz; 14.1005-14.112 MHz; 18.105-18.110 MHz; 21.090-21.100 MHz; 24.925-24.930 MHz; and 28.120-28.189 MHz.

Manually controlled stations may initiate communications with automatically controlled HF RTTY and data stations on any frequency where such emissions are authorized, but the automatically controlled station's emissions may not occupy a bandwidth larger than 500 Hz [97.221(b)].

Q&A—Data, RTTY Modulation Methods

Q. Is it permissible to modulate an FM voice transmitter with data or RTTY?

A. It depends on the frequency. It is permissible only where data or RTTY emissions are authorized above 50 MHz. This type of transmission is commonly used in the VHF bands because of the ready availability of FM transceivers.

Some amateurs have expressed a desire to operate using this mode in the 10-meter band above 29.0 MHz. Although FM voice is specifically authorized as a mode, neither data nor RTTY is permitted on 10 meters, except in the 28.0-28.3 MHz segment [97.305(c); 97.307(f)(1)].

Q. Are there any speed limits or bandwidth constraints for data or RTTY?

A. Yes, at least below 450 MHz. The limits break down as follows [97.307(f)]:

Frequency Range	Speed Limit for Specified Codes	Maximum Bandwidth for Unspecified Codes
1.80 - 24.93 MHz	300 bauds	Not permitted
28.0 - 28.3 MHz	1200 bauds	Not permitted
50.1 - 148 MHz	19.6 kilobauds	20 kHz
222 - 450 MHz	56 kilobauds	100 kHz
Above 902 MHz	No speed limit	No bandwidth limit

The maximum bandwidth is, in this case, the "width of a frequency band outside of which the mean power of the transmitted signal is attenuated at least 26 dB below the mean power of the transmitted signal within the band" [97.3(a)(8)]. Where no speed or bandwidth limits are listed, the basic rule is to stay within the ham band. There are, however, some operational considerations. The HF RTTY/data bands are crowded, and it wouldn't set too well with other hams if you were to use excessive bandwidth. The practical limitation at HF is imposed by the audio characteristics of the SSB transceivers generally used. Typically, they may have −6 dB audio bandwidths of 2.1 to 2.7 kHz. So packet radio bandwidths on the HF bands have been limited to about 2 kHz. For example, the packet frequencies just above 14.100 MHz are spaced every 2 kHz. In all cases, check the bandplan, and where appropriate, contact your frequency coordinator to determine the appropriate bandwidth for your operation.

Q. Can I use a parallel modem on HF?

A. This refers to a modem that uses several tones, rather than switching between two, and keys them at a slower rate to avoid the effects of multipath (intersymbol) distortion. The emission J2D does not limit the number of tones; it simply limits the transmission to a single channel of information. So, the answer is yes, if you use J2D emission.

CHAPTER 14

Serving the Public

Emergency Communications: Are They Legal?

All *emergency* communications are legal. Yet, because of folklore and second-hand anecdotes, emergency and public-service communications have become confusing for many amateurs. A surprising number seem to believe that much amateur public-service and disaster communications are illegal. Most of this confusion can be traced to the FCC's Rules banning any form of "business communications," which were in effect from 1971 until 1993, as discussed below.

Some Examples

During August and September 1987, nearly *700 square miles* of California forest lands were destroyed by wildfire, causing the evacuation of tens of thousands of mountain residents. Hundreds of ham operators provided support communications for the US Forest Service, the California Department of Forestry, the American Red Cross and other relief agencies.

Once the fires were out, several hams were heard asking, "Were we legal? Or, were we conducting the regular business of these relief agencies?" That this question was asked at all, under the circumstances, illustrates the confusing interpretations of the FCC Rules within the amateur fraternity.

In some instances, the misunderstandings about emergency communications have tarnished the image of Amateur Radio. At a recent 200-mile bike ride, a "sag wagon" with Amateur Radio communications arrived on the scene of a serious accident; a volunteer paramedic was already present and administering first aid.

Because of the extent of the injuries, the paramedic asked to confer with a physician who happened to be in the vicinity of the amateur net control station. Strangely, the net control operator refused to allow the physician to speak directly over the radio. In spite of complicated medical terminology and the potential for mistakes, the net control operator insisted on verbally relaying each message. The control operator said he wasn't sure if it would be legal for the paramedic to speak directly with the physician.

Unfortunately, this paramedic is a volunteer with a search and rescue group that *needs* Amateur Radio support. But, based on this unfortunate incident, they have chosen to avoid ham radio because, the paramedic said, "It's unreliable."

At another bike tour, open to the public and sponsored by a local bike club, several hams told the ham radio coordinator that helping the bike club was not legal for ham radio—yet seven of the 2500 riders suffered major injuries requiring paramedic or helicopter air ambulance response. Ham radio proved essential to the safety of the riders.

When Does a Situation Become an Emergency?

Communications "in connection with the immediate safety of human life and immediate protection of property when normal communication systems are not available" are legal [97.403]. The difficulty is the interpretation of what constitutes an immediate threat to life or property.

Obvious examples of an emergency include natural disasters—such as tornadoes, hurricanes, blizzards, floods—and other forms of severe weather, forest fires, landslides and earthquakes. These typically cause immediate danger to life and property and outages of normal communication systems (telephones and public-safety radio systems).

During an emergency, you may use your radio in any appropriate manner. Even though putting out fires or providing disaster assistance may be the *regular* business of your fire department or of the American Red Cross, in these situations an emergency affecting the immediate safety of life and property has occurred, and your Amateur Radio participation is not only allowed, but encouraged. Under these guidelines, assisting the Forest Service during a wildfire, allowing a physician to use your radio or performing Red Cross disaster assessment are all legitimate Amateur Radio operations.

Other situations, however, are less clear cut. For example, you spot a motorist, stranded along a suburban highway. Can you call for a tow truck on the repeater autopatch? Under the old rules this could have been construed as "business com-

The Boston Marathon: When a seriously injured runner was on the way for needed medical attention, Kevin Erickson, N1ERS, in the communication trailer in downtown Boston helped relay messages from the "sweep" bus team to the medical facility at the finish line. Some runners picked up by the buses experienced heat illness, dehydration, hypothermia and foot injuries. *(Photo courtesy NQ1R)*

munications"; under the current rules it's perfectly okay. Requesting police assistance for a stranded motorist would have been legal at any time. While this example may hardly seem like an emergency, it represents a real danger to the stranded motorist. In 1988, a San Francisco Bay Area mother and daughter were killed when their disabled automobile was struck from behind. They were parked well off the right side of the freeway, emergency flashers on, hood up, in broad daylight. So pick up your H-T and call!

At the scene of an accident, can you hand your radio to an unlicensed person, such as a fire chief? Yes; as long as you remain the control operator, this is merely standard third-party operation [97.115(b)(1)]. In fact, this is usually the most efficient way to provide emergency communications to an agency. Instead of relaying the message yourself, why not put the sender and the recipient on the radio? This eliminates errors and is much more efficient.

Extraordinary Communications

If you have equipment capable of operating beyond amateur privileges, and you are present when an emergency occurs or find yourself in extreme distress, can you use that ability to call for help? Under §§97.403 and 97.405, the answer is a qualified yes. §97.403 states that "No provision of these rules prevents the use by an amateur station of any means of radiocommunication at its disposal to provide essential communication needs in connection with the immediate safety of human life and immediate protection of property when normal communications systems are not available."

Note the words "essential" and "immediate," and the phrase "when normal communications systems are not available." If you can possibly communicate by any other method, including non-radio means, do so. If the situation doesn't in-

On January 17, 1994, a major earthquake centered in the Los Angeles suburb of Northridge flattened buildings, destroyed highways and disrupted communications. Amateurs came through, providing essential early communications. Although his car was crushed, K6GF escaped uninjured from this collapsed apartment complex. (Photo by Jeff Reinhardt, KM6II)

volve an immediate threat to life or property, or if the communications are not essential to relieve the situation, keep quiet! If you do operate on frequencies assigned to another service, remember that you must keep any communications to the absolute minimum to pass the information. Once the information is passed, stay off the channel. Public service agencies do not appreciate unauthorized people operating on their frequencies, so don't be surprised if you encounter resistance. Be prepared to justify your actions afterward.

These rules are intended to let you know that if you happen to have the capability, under extreme conditions you may use it. You should not intentionally provide for such operation. If you feel you must prepare for emergency communications beyond what can be handled by amateur operations, you are expected to do so using legitimate services.

Public Events Communication

The opening of this chapter discussed concerns about the appropriateness of some public service communications. FCC licensees, amateur or otherwise, are supposed to serve "the public interest, convenience, and necessity." In §97.1 of the Commission's Rules, the basis and purpose of the Amateur Radio Service is spelled out clearly. Obviously, public service and educational activities are to be actively encouraged; how could there be any question about their being legal?

Around 1970 there were concerns about possible abuses of Amateur Radio by non-amateur and business interests. These concerns led the FCC to prohibit amateur communications "to facilitate the business or commercial affairs of any party" or "as an alternative to other authorized radio services." Over time, the interpretations of these rules became progressively more literal until they had a chilling effect even on meritorious public service activities. Something had to be done to put things back on track.

On September 13, 1993, following a rulemaking proceeding (PR Docket 92-136), the FCC dropped the old "no business" language, and replaced it with a prohibition on communications for compensation, on behalf of one's employer, or in which the amateur has a pecuniary interest [97.113(a)(2),(3)]. In place of the flat prohibition on providing an alternative to other radio services is a less restrictive one against doing so on a regular basis [97.113(a)(5)].

These changes mean a lot to public service-oriented amateurs. They remove the ambiguities that have plagued amateur public-service communications for the past two decades, and have curtailed the endless hair-splitting discussions about whether particular communications were permitted.

The focus now is on whether the amateur, or his or her employer, stands to benefit financially, rather than on the content of the communication [97.113(a)(2),(3)]. If so, then the communication is still prohibited. If not, then the remaining question is whether the communications need is one that ought to be met by some other radio service. Here, the rule of reason applies. A need that arises on a regular basis, and for which other communications services are reasonably available, should not be met by Amateur Radio. The FCC declined to define "regular," but this shouldn't pose much of a problem for us since abuses will tend to be self-limiting; volunteers don't like being taken advantage of, and if they are they should just say no. One popular activity for which there is no practical communications alternative available, collecting data for the National Weather Service, was singled out by the FCC as an example of what is permitted under the new rules [Report and Order to PR Docket 92-136, 97.113(a)(5)].

The new rules do not represent a philosophical departure from our "roots." In fact, they are almost identical with the regulations in effect prior to the "no business communications" rule. They provide latitude in our operating and especially in our public-service communications, just as we had for decades before the onset of over-regulation in the early '70s. This is one of those rare times when we get to return to the "good old days."

Tactical Call Signs?

Tactical call signs are often used when working with other agencies during an emergency, or during large public-service activities. For example, during a running race, names like "Finish line," "Mile 1," "Mile 2," "First Aid 1" and "Water truck" quickly identify each function and eliminate confusion when working with other agencies, such as a fire department, where amateur call signs are meaningless. They also help prevent confusion when several operators may take turns at a position.

The use of tactical call signs is a good idea, but it in no way relieves you of the obligation to identify your operation under the FCC's Rules for normal station identification. You must still give your FCC-assigned call sign at the end of your communication, and at least every 10 minutes during the contact [97.119].

Can You be Compensated for Your Amateur Radio Assistance?

No. The FCC regulations prohibit payments for the use of an Amateur Radio station [97.113(a)(2)]. This rule does not prohibit you from being reimbursed for incidental expenses unrelated to your radio communication. If you assist at a disaster scene 100 miles from your home, you are not prohibited from receiving reimbursement for out-of-pocket travel expenses unrelated to your radio communication. For example, if as an American Red Cross Disaster Services volunteer, you are flown to the scene of a disaster where you happen to use Amateur Radio as part of the relief effort, you are not required to pay your own airfare.

On the other hand, if your employer sends you to a disaster area, or encourages you to participate in disaster relief by paying your wages while you are doing so, then you may not engage in Amateur Radio communications since you are being compensated for your activities. As part of disaster preparedness emergency stations may be established at hospitals, pharmaceutical companies, utilities, and the like. These stations will probably be involved in drills, and may be involved in actual disaster communications. If you are employed at one of these places, you cannot participate in such communications while you are "on the clock." It can get very sticky if you are expected to operate such a station during an actual emergency, so guidelines for manning the station need to be worked out in advance [97.113(a)(3)]. Remember, however, that the key words defining an emergency are: "immediate safety of human life and immediate protection of property when normal communication systems are not available." When the chips are down, you may do whatever is necessary to get the message through.

Disaster Communications

If normal communications are down or overloaded because of a pre- or post-disaster situation, hams may provide essential communications needs to facilitate relief actions [97.401(a)]. If you're outside an area regulated by the FCC, you may still perform these functions under the international Radio Regulations [97.401(b)].

If an emergency strikes a widespread area, disrupting the normal lines of communication, the FCC Engineer-In-Charge of the area may designate certain frequencies for use by stations assisting the stricken area only [97.401(c)]. All amateur transmissions with, or within, the designated area conducted on the FCC-designated emergency frequencies must pertain directly to relief work, emergency service or the establishment and maintenance of efficient networks for handling emergency traffic. The FCC may also set forth further special conditions and rules during the communications emergency [97.401(c)].

Also, the Commission may designate certain amateur stations to police the emergency frequencies, warning noncomplying stations operating there. The emergency conditions imposed by the FCC can be lifted only by the FCC or its authorized representative. Amateurs desiring a declaration of a communication emergency should contact the FCC Engineer-In-Charge of the area concerned [97.401(c)].

RACES

Founded in 1952 with the help of the ARRL, the Radio Amateur Civil Emergency Service is sponsored by local and state civil defense organizations and supported by the Federal Emergency Management Administration (FEMA). RACES works principally at the local level, through local and state civil defense agencies

---

Congress on Amateur Radio

One of the last pieces of business accomplished by the US Congress in 1988 was a resolution expressing its feelings about the volunteer public-service work of radio amateurs:

SENSE OF CONGRESS

The Congress finds that—

(1) More than 435,000 radio amateurs in the United States are licensed by the Federal Communications Commission upon examination in radio regulations, technical principles and the international Morse code;

(2) By international treaty and the Federal Communications Commission regulations, the amateur is authorized to operate his or her station in a radio service in intercommunications and technical investigations solely with a personal aim and without pecuniary interest;

(3) Among the basic purposes for the Amateur Radio service is the provision of voluntary, noncommercial radio service, particularly emergency communications; and

(4) Volunteer Amateur Radio emergency communications services have consistently and reliably been provided before, during and after floods, tornadoes, forest fires, earthquakes, blizzards, train wrecks, chemical spills and other disasters.

It is the sense of the Congress that—

(1) It strongly encourages and supports the Amateur Radio service and its emergency communications efforts; and

(2) Government agencies shall take into account the valuable contributions made by Amateur Radio operators when considering actions affecting the Amateur Radio service.

---

organized by state governments, to provide emergency communications when activated by the appropriate civil defense authority. Originally it had its own section of the FCC Rules; now it's incorporated in Subpart E of the amateur rules.

RACES is intended to provide radio communications for civil defense purposes only, during periods of local, regional or national civil emergencies. These emergencies can include war-related activities, but are more likely to involve natural disasters such as fires, floods and earthquakes. RACES operation is authorized only by the appropriate local, state, or federal official, and is strictly limited to official civil defense activities in an emergency communications situation.

One important aspect of RACES is that, when activated, it can continue in operation even if normal amateur operation is suspended during a national emergency, such as a declaration of war. Should amateur operation be suspended, RACES operation would only be permitted on specific frequencies; see the accompanying sidebar.

Prior to 1990, a war or national emergency proclamation by the President automatically mandated the closing of all Amateur Radio stations. Chapter II of 47 CFR (which deals with the National Security Council) was changed on December 11, 1990, to eliminate that requirement. Instead, when a national emergency is

---

Good Amateur Practice in Disasters: A Case Study

During Hurricane Gilbert, which devastated the island of Jamaica in the Caribbean and wiped out its communications infrastructure, amateurs were the first to get word out and performed heroically providing essential communications. However, we also learned a number of lessons that should help us provide better communications in future disaster situations:

• Communications channels in support of damage assessment and disaster relief agencies were slow to be established, and in some cases had to be established outside the amateur bands. (Incredibly, some stations even followed these officially designated stations outside the bands, without authorization!)

• Preoccupation with welfare inquiry traffic got in the way of higher priority communication. The Red Cross placed a moratorium on inquiries to Jamaica for several days; this provided time for outbound welfare messages to be sent, each of which could potentially head off at least one, and perhaps many, inquiries. Accepting inquiry traffic before there is any way to handle it raises false hopes and clogs the system.

• In a disaster, control belongs in the affected area; people there are in the best position to know the priorities. Those of us fortunate enough to be outside the disaster area are there to support them, not the other way around.

• In providing information to the media, amateurs must make sure what is being passed along is authentic, and not unsubstantiated hearsay.

General Considerations In Emergencies

• Use your receiver more, your transmitter less. The tendency to transmit rather than listen causes excessive QRM.
• Monitor emergency net frequencies.
• Listen to W1AW for the latest bulletin and news.

RACES Frequencies

All of the authorized frequencies and emissions allocated to the Amateur Radio Service are also available to RACES on a shared basis [97.407(b)]. If Amateur Radio operation is suspended by a Presidential proclamation of national emergency, RACES, if activated, may operate only on these frequencies. There are specific operating limitations with respect to these frequencies, and in all cases not specifically covered by the RACES rules, amateurs engaging in RACES operation are governed by the provisions of the rules governing Amateur Radio stations and operators.

RACES Frequencies

kHz	MHz	
1800-1825	10.10-10.15	52-54
1975-2000	14.047-14.053	50.350-50.750
3500-3550	14.220-14.230	144.50-145.71
3930-3980	14.331-14.350	146-148
3984-4000	21.047-21.053	222-225
7079-7125	21.228-21.267	420-450
7245-7255	28.550-28.750	1240-1300
	29.237-29.273	2390-2450
	29.450-29.650	

declared, amateurs are obligated to observe whatever orders the FCC may issue in the interests of national security.

Operation in RACES

The FCC Rules permit two types of stations to operate as part of RACES. These are FCC licensed RACES stations, and amateur stations that have been properly registered with a civil defense organization. The FCC stopped issuing new RACES station licenses in 1978, but continues to renew existing licenses. RACES stations may operate only during declared emergencies, or during drills ordered by the served civil defense organization. The control operator of a RACES station or an amateur station registered in RACES must be a person holding an FCC-granted amateur operator license, and is bound by the privileges authorized by their license.

Operation in RACES is highly regulated. Amateur stations operating in RACES must be registered with a specific civil defense organization, and may only communicate with other stations registered with the same body unless specifically authorized to communicate with other RACES stations by the local civil defense authority. No station operating in RACES may communicate with a station that is not operating in RACES except, when authorized, another US Government station or an FCC licensed station in another service.

RACES Communications

The types of communications permitted are similarly restricted. Aside from drills, communications in RACES may consist only of civil defense messages concerning:

1) public safety or national defense during times of local, regional, or national emergencies;
2) immediate safety of life or property, the maintenance of law and order, alleviation of human suffering and need, and the combating of armed attack or sabotage;
3) the accumulation and dissemination of public information or instructions to the civilian population as required by the civil defense organization or other government agencies or relief agencies.

RACES drills must be authorized by the responsible civil defense official, and may not exceed a total time of one hour per week. With approval of the chief emergency planning officer at the state, district or territory level, drills of up to 72 hours duration may be conducted no more than twice a year.

RACES and ARES

Because of these restrictions, RACES cannot be activated to perform non-emergency public service activities such as assisting in a race or parade. This is where ARES, the Amateur Radio Emergency Service comes in. Unlike RACES, which is a separate regulated service sharing spectrum space with the Amateur Service, ARES is part of the Amateur Service and subject to all its regulations and privileges. Because it has no statutory status, in the unlikely event of suspension of amateur operation, ARES must shut down. On the other hand, where there is no local civil defense organization, required to support RACES activity, amateurs can organize under ARES to train and provide emergency services. Because of the limitation on drills, many RACES operations work with the ARES organization to train for disasters. Where RACES cannot act in nonemergency situations to provide public service communications, ARES can. The two operations can and should compliment each other, but be aware that they are very different animals.

CHAPTER 15

Regulation: The Big Picture

The electromagnetic spectrum is a limited resource; each kilohertz is precious to the competing interests that lay claim to it. Fortunately, unlike natural resources, the spectrum is a limited but nondepletable resource that, if misused, can be restored to normal as soon as the misuse stops. Each day gives us a new opportunity to use the spectrum efficiently and intelligently.

Who decides where Amateur Radio frequencies will be, or, for that matter, where your favorite AM or FM broadcast station will be? With the proliferation of competing services within the physical confines of the radio spectrum comes the need for controls. Without such controls, chaos would reign with radio services colliding with each other. Not surprisingly, controls have developed over the years, implemented at the international and domestic levels. The International Telecommunication Union (ITU) has the vital role of dividing up the range of communications frequencies for the entire world. Based on the demonstrated or perceived needs of different services, which include commercial broadcast, land mobile and private radio (including Amateur Radio), member-nations of the ITU decide which radio services will be given which band of frequencies. Once that's done, the government regulators for each particular country take over.

International Regulation of the Spectrum

Nations sign all sorts of treaties and agreements (international communications, nuclear arms, ozone depletion and so on) to bring some order to international relationships. Without these agreements, anarchy would prevail. Those affecting Amateur Radio have proved quite effective over the years. These must be *international* in scope because radio waves know no geographical or political boundaries—they don't stop for customs inspections.

Amateur Radio frequency band allocations don't just happen. Band allocation proposals must survive a maze of national agencies, the ITU and the gauntlet of other spectrum users. The ITU allocates portions of the radio spectrum to each service vying for its own slice of the pie. The ITU develops regulations designed to reduce potential interference problems.

The process occurs at the international conference table. Member-nations' delegates bring their countries' official positions to the table. They debate the merits of these positions, taking into consideration changes in technologies since the previous conference and the needs of each administration. They finally arrive at an International Table of Allocations and Radio Regulations. Amateur Radio is provided for in Article 32 of those regulations.

The international Radio Regulations of the ITU affect all radio amateurs. The US has a responsibility to make rules that are consistent with these international agreements. Once the international allocations have been decided, it's up to the Federal Communications Commission (FCC) to decide the best way to allocate frequency bands to those services using them in the US. The FCC is the governing body in the US when it comes to Amateur Radio. The Communications Act of 1934, as amended, is the chief tool by which the US carries out its obligations to the world's telecommunications community. Through the authority delegated to the FCC by Congress in the Communications Act, the FCC adopted a body of rules to deal with communications. Part 97 (officially cited as Part 97 of Title 47 CFR [Code of Federal Regulations]) of the FCC Rules governs Amateur Radio.

Federal Communications Commission

What is the FCC? It's the outfit that issues your license, for one thing! The FCC is the US government agency charged with regulating interstate and foreign communication involving radio, television, wire cable and satellites. The FCC does not function like any other government department; it is a sovereign federal agency created by Congress and, as such, reports directly to Congress.

The FCC allocates bands of frequencies to nongovernment communications services and assigns frequencies to individual stations. It licenses and regulates stations and operators, and regulates common carriers in interstate and foreign communications by telegraph, telephone and satellite. One job it does *not* do is regulate federal government radio operations; this is done by the National Telecommunications and Information Administration (NTIA), under delegated authority of the President.

The FCC consists of five commissioners appointed by the President with the approval of the Senate. No commissioner can have a financial interest in any FCC-regulated business. Appointments are for five years, except when filling an unexpired term. One of the commissioners is appointed to be chairman by the President, and his tenure as chairman runs with the President's term of office. The chairman may revert to commissioner if the President leaves office before the end of the chairman's term as commissioner. As with many federal agencies, the commissioners function "collegially" (that is, one man—one vote), supervising all FCC activities, with substantial delegation of responsibilities to FCC staff members. The chairman is responsible for the overall administration of the internal affairs of the FCC and sets the basic agenda for the agency. Otherwise, as noted above, he, like his fellow commissioners, has only one vote in formal policy decisions. The internal structure of the FCC is shown in Fig 1.

Policy determinations are made by the commission as a whole. FCC practices conform to the Communications Act of 1934, as amended, the Administrative Procedure Act, as amended, and other applicable laws.

The FCC cooperates with other agencies, such as those involved with radio and wire communication in international and domestic matters. It also cooperates with radio-user groups, such as the ARRL.

FCC regulation of radio includes consideration of applications for construction permits and license for all classes of nongovernment stations, and frequency, power and call sign assignments. The FCC is involved in authorization of communications circuits, modification and renewal of licenses, and inspection of transmitting equipment and regulation of its use. The FCC has enforcement powers and

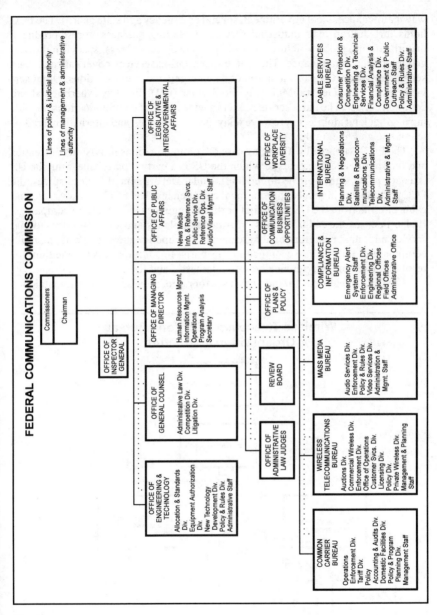

FEDERAL COMMUNICATIONS COMMISSION

— Lines of policy & judicial authority
···· Lines of management & administrative authority

Commissioners
Chairman

OFFICE OF INSPECTOR GENERAL

OFFICE OF LEGISLATIVE & INTERGOVERNMENTAL AFFAIRS

OFFICE OF PUBLIC AFFAIRS
News Media
Info. & Reference Svcs.
Public Service Div.
Reference Ops. Div.
Audio/Visual Mgmt. Staff

OFFICE OF MANAGING DIRECTOR
Human Resources Mgmt.
Information Mgmt.
Operations
Program Analysis
Secretary

OFFICE OF GENERAL COUNSEL
Administrative Law Div.
Competition Div.
Litigation Div.

OFFICE OF ENGINEERING & TECHNOLOGY
Allocation & Standards Div.
Equipment Authorization Div.
New Technology Development Div.
Policy & Rules Div.
Administrative Staff

OFFICE OF WORKPLACE DIVERSITY

OFFICE OF COMMUNICATION BUSINESS OPPORTUNITIES

OFFICE OF PLANS & POLICY

REVIEW BOARD

OFFICE OF ADMINISTRATIVE LAW JUDGES

CABLE SERVICES BUREAU
Consumer Protection & Competition Div.
Engineering & Technical Services Div.
Financial Analysis & Compliance Div.
Government & Public Outreach Staff
Policy & Rules Div.
Administrative Staff

INTERNATIONAL BUREAU
Planning & Negotiations Div.
Satellite & Radiocommunications Div.
Telecommunications Div.
Administrative & Mgmt. Staff

COMPLIANCE & INFORMATION BUREAU
Emergency Alert System Staff
Enforcement Div.
Engineering Div.
Regional Offices
Field Offices
Administrative Office

MASS MEDIA BUREAU
Audio Services Div.
Enforcement Div.
Policy & Rules Div.
Video Services Div.
Administration & Mgmt. Staff

WIRELESS TELECOMMUNICATIONS BUREAU
Auctions Div.
Commercial Wireless Div.
Enforcement Div.
Office of Operations
Customer Svcs. Div.
Licensing Div.
Policy Div.
Private Wireless Div.
Management & Planning Staff

COMMON CARRIER BUREAU
Operations
Enforcement Div.
Tariff Div.
Policy
Accounting & Audits Div.
Domestic Facilities Div.
Policy & Program Planning Staff
Management Staff

Fig 1

may levy fines in cases of noncompliance with its rules. In sum, the FCC carries out the Communications Act.

The FCC has authority over the Amateur Radio Service, but it mainly concerns itself with a number of other services. These include aviation (aircraft and ground); marine (ship and coastal); public safety (police, fire, forestry conservation, highway maintenance, local government, special emergency and state guard). The FCC holds jurisdiction over industrial services including business, forest products, manufacturers, motion pictures, petroleum, power, relay press, special industrial and telephone maintenance. The land transportation services cover railroad, passenger and truck, taxicab and automobile emergency communications. Other services include the General Mobile Radio Service (CB), disaster, experimental and common carrier. The latter service covers a wide range of communications, including paging, land mobile, microwave relay, broadcast relay, and international radiotelephone and telegraph.

The FCC is charged with taking care of communications provisions of treaties and international agreements to which the US is a party ("party" meaning the US is a nation that signed on the dotted line). Under Department of State auspices, the FCC participates in related international conferences. The FCC licenses radio and cable circuits from the US to foreign ports and regulates the operating companies. It also licenses the radio stations on American planes and ships in international service and, under international agreements and upon request, inspects the radio equipment of foreign vessels touching US ports. In addition, it is the medium for resolving cases of interference between domestic and foreign radio stations.

The FCC is required "to study new uses for radio, provide for experimental uses

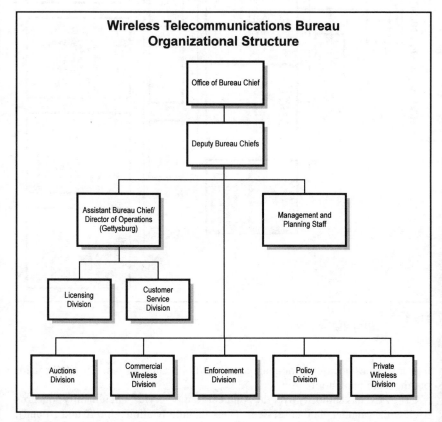

Fig 2

of frequencies, and generally encourage the larger and more effective use of radio in the public interest." Cooperation is maintained with government and commercial research activities. The FCC operates a laboratory at Laurel, Maryland. It also carries out studies to provide information on complex questions facing the FCC and the telecommunications industry as a whole.

Wire and radio communications facilities that are used to aid national defense form one of the basic requirements of the Communications Act. The President has delegated some of these functions to the FCC.

The functions relating to assignment of frequencies to radio stations belonging to, and operated by, the US government were assigned to the Assistant Secretary of Commerce for Communications and Information (who also holds the title of NTIA Administrator).

Internal Structure of the FCC

In the fall of 1994, the FCC underwent a major reorganization. The new Wireless Telecommunications Bureau (WTB) included functions of the old Private Radio Bureau. See Fig 2. An International Bureau and a Cable Services Bureau were also created. According to the FCC, personal communications systems and other emerging technologies made the WTB necessary, and the new International Bureau "will better meet the challenges ahead as the FCC continues its role in international telecommunications" (FCC Chairman Reed Hundt).

The FCC staff is organized on a functional basis. There are now six operating bureaus: Mass Media, Common Carrier, Compliance and Information (formerly the Field Operations Bureau), International, Cable Services, and Wireless Telecommunications. In addition there are 11 staff offices: Managing Director, Engineering and Technology, Public Affairs, Plans and Policy, General Counsel, Administrative Law Judges, Legislative Affairs, Inspector General, Small Business Activities, Workplace Diversity and the Review Board. FCC headquarters are in Washington, DC.

The FCC field staff is in field offices throughout the country. The field staff engages for the most part in engineering work. This includes monitoring the radio spectrum to see that station operations meet technical requirements, inspecting stations of all types and issuing permits or licenses to those found qualified. It locates and closes unauthorized transmitters, furnishes radio bearings for aircraft or ships in distress, locates sources of interference and suggests remedial measures. The field staff performs special engineering work for other government agencies, and obtains and analyzes technical data for FCC use.

Where does Amateur Radio fit into the FCC's scheme of things? Amateur Radio is administered by the Wireless Telecommunications Bureau, specifically the WTB's Private Wireless Division. Amateur Radio is considered private because it is a two-way service for individuals to use for private, strictly noncommercial communications. In contrast, broadcasting serves a mass audience and common-carrier services (such as the telephone system) provide a communications service for hire.

Included under the Wireless Telecommunications Bureau heading is the Licensing Division in Gettysburg. Other components of the FCC hierarchy of concern to Amateur Radio are the Office of Engineering, Science and Technology and the Compliance and Information Bureau (Regional Offices, Field Offices and Monitoring Facilities).

Deregulation

In the 1980's, a mandate was passed down to regulatory agencies that there should be less intrusion into people's lives by the federal government, with the understanding that a marketplace philosophy—based on individual initiative and free enterprise—should prevail whenever possible. The enactment of the Communications Amendments Act of 1982 (Public Law 97-259) opened the door to sweeping changes in the amateur service and placed the primary responsibility for protecting the future of Amateur Radio where it belongs: On amateurs themselves. Signed into law by President Reagan on September 14, 1982, PL 97-259 amended the Communications Act in several critical areas (all of which have been incorporated into Part 97), as follows:

- Authority was vested in the FCC to regulate the susceptibility of electronic equipment to RFI.
- The amateur service was exempted from the "secrecy of communications" provisions of §705 of the Communications Act. This cleared the way for a more active role on the part of amateurs to help regulate their own bands.
- Because of this exemption, the FCC was legally authorized to use volunteers to monitor the bands for rules violations and convey information to FCC personnel. This formed the basis for the creation of the League's Amateur Auxiliary to the FCC's Compliance and Information Bureau.
- The FCC was authorized to use volunteers in preparing and administering exams, which led the League (and others) to become Volunteer Examiner Coordinators and coordinate exams throughout the US and overseas.
- The term of an amateur license was increased from 5 to 10 years to reduce the administrative burden on the FCC and amateur licensees.

Most observers agree that telecommunications deregulation has generally been positive. The League has, for the most part, supported the FCC's deregulatory actions in Amateur Radio because amateurs have demonstrated time and again the ability to keep their own shop in order. In recent years, the FCC has, among other actions, eliminated mandatory logging, simplified station ID, permitted ASCII, AMTOR, packet and other digital codes, expanded Novice privileges and established the codeless Technician license.

How FCC Rules are Made

The FCC Rules are not just handed down to us from the Commission; amateurs have a right to directly participate in the rule making procedure. As American citizens, we can and do have a profound effect on what rules are added, dropped or modified. See Fig 3.

The Administrative Procedure Act

The Administrative Procedure Act and certain sections of the Communications Act set forth specific procedures that all administrative agencies must follow in adopting and amending their rules. The Act also sets forth the procedures to be followed in adjudicatory (that is, trial-like) hearings.

Where rule making is concerned, the essential provisions of the Act are (1) Public notice of the proposal in the *Federal Register*, and (2) The right of interested persons to submit written comments. This is "notice-and-comment" rule making. Prior notice need not be given if an agency, for good cause, finds that the notice

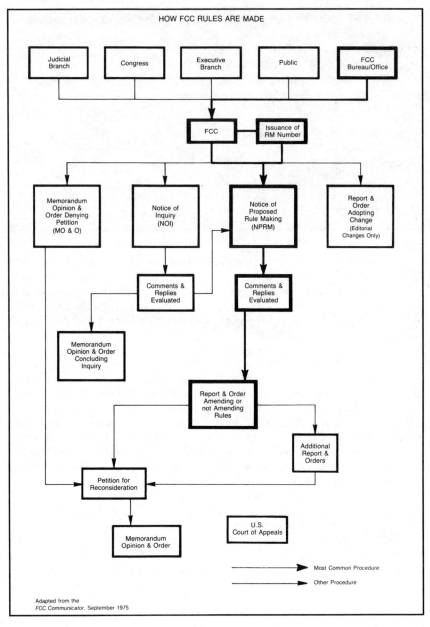

Fig 3

and comments are impractical, unnecessary or contrary to the public interest. Overall, rules may be adopted, amended or repealed by an agency on its own initiative, or may be requested by any interested person who files a petition for rule making. If the FCC feels a petition has merit, a rule making (RM) file number is assigned.

```
┌─────────────────────────────────────────────────────────────────────┐
│  Caption Format to be Used for Petition or Formal Comments            │
│                                                                       │
│   Before the                                                          │
│   FEDERAL COMMUNICATIONS COMMISSION                                   │
│   Washington, DC 20554                                                │
│                                                                       │
│   In the Matter of                      )                             │
│                                         )                             │
│   Amendment of Section 97.–             )                             │
│   (Insert rule number if an             )        RM-    (or docket number) │
│      amendment is                       )                             │
│      proposed, and the                  )                             │
│      subject matter of the              )                             │
│      Amateur Radio Service              )                             │
│      Rule(s).)                          )                             │
│                                                                       │
└─────────────────────────────────────────────────────────────────────┘
```

Some petitions that plainly do not warrant FCC consideration are not given RM file numbers and are usually dismissed by an FCC bureau chief under delegated authority. Any interested person may file a statement in support of or in opposition to a petition for rule making not later than 30 days from issuance of the Public Notice. Replies to supporting or opposing statements are due not later than 15 days after the filing of such statements.

Notice of Proposed Rule Making

Rule making involving Amateur Radio matters usually falls under the jurisdiction of the FCC's Wireless Telecommunications Bureau. Petitions involving amateur matters are processed by the Private Wireless Division of the WTB. If a petition has merit, a draft Notice of Proposed Rule Making (NPRM) may be submitted by the WTB Chief to the Commissioners for their consideration. If adopted for release—which requires a majority vote by the Commissioners—a Docket Number is assigned (PR Docket No. 88-139, for example). The entire NPRM is then released to the public and published in the *Federal Register* for comments.

Depending on the FCC's workload, its priorities and the amount of interest shown by concerned parties, several months or even years may elapse between the release of an NPRM and when the FCC reaches a final decision, if at all.

The Administrative Procedure Act requires that the NPRM set forth the terms or substance of the proposed rule or a description of the issues involved, and a reference to the legal authority for the rule making. It also requires a statement of the time, place and nature of public rule making proceedings. Deadlines for filing comments and reply comments are given (reply comments are those filed in response *not* to the underlying NPRM but to the *comments that were filed* in response to the underlying NPRM which are available to the public at the FCC's public reference room in Washington). The NPRM must be published in the *Federal Register*. Interested persons have an opportunity to participate in the proceedings through the submission of written data, views or arguments.

If you are filing comments in an NPRM, use the caption format shown above, making sure that you put the appropriate docket number at the top.

When you write your comments, list your credentials, be fact-specific and

Hints on Filing Comments with the FCC

The FCC is interested in any experiences, knowledge, or insights that outside parties may have to shed light on issues and questions raised in the rule making process. The public and industry have the opportunity to comment upon Petitions for Rule Makings, NOI's, NPRM's, Further NPRM's, Report and Orders, and others' comments on the aforementioned documents. It is a common misconception that one must be a lawyer to be able to file comments with the FCC, but all that is necessary is an interest in an issue and the ability to read and follow directions.

Prior to drafting comments, it is crucial to read and understand fully the item you wish to comment on. Usually, the NPRM, NOI, or other item will specify and invite comment upon the issue(s) that the Commission is interested in studying further. Examination of the issue(s) and relevant documents is the most important part of the comment process. Comments may take any form, but below are some hints to assist you in writing them.

Format: There is no required format for informal comments, although if you plan to file formally, it is required that they be typed, double-spaced, and on 8.5" × 11" paper. Additional requirements for formal filings are set forth in Sections 1.49 and 1.419 of the FCC Rules. The Docket Number or Rule Making Number of the item at hand should be included on your comments, and can be found on the front page of the Commission document or public notice. You should also include your name and complete mailing address.

Content: Your comments should state who you are and what your specific interest is. (You do not need to represent yourself in an official capacity. You may, for example, express your opinion as a concerned consumer, concerned parent, etc. and sign your name.) State your position and the facts directly, as thoroughly but as briefly as possible. Explain your position as it relates to your experience and be explicit. Make clear if the details of a proposed rule or only one of several provisions of the rule are objectionable. If the rule would be acceptable with certain safeguards, explain them and why they are necessary.

Support: Statements of agreement or dissent in comments should be supported to the best extent possible by factual (studies, statistics, etc.), logical, and/or legal information. Support should illustrate why your position is in the public interest. The more support made, the more persuasive the comments will be.

Length: Comments may be any length, although it is preferred that they be succinct and direct. If formal comments are longer than ten pages, it is required that they include a summary sheet.

Time Frame: Your comments should be submitted well within the time frame designated on the original document or public notice. It is almost always included on the first page of an NPRM or NOI. However if the deadline has passed, you can still submit your views informally in a permissible ex parte presentation.

Filing: Send your written comments to Secretary, Federal Communications Commission, 1919 M Street, N.W., Washington, D.C. 20554. If you wish your comments to be received as an informal filing, submit the original and one copy. If you want your comments to be received as a formal filing, you should submit an original and four copies. For more specific filing information, please refer to the FCC Public Notice

to the point. When you comment on an active NPRM, the FCC wants you to send in an original and four copies, for a total of five documents (make an additional copy for your own records). If you want to be sure that each FCC commissioner gets a personal copy, send five additional copies. So, if you want to have all bases covered, your envelope to the FCC must contain 10 documents, the original and nine copies. To be considered formally, comments must have type at least 12 points in height, be double spaced and be printed on $8^1/_2 \times 11$-inch paper with the printed material not exceeding $6^1/_2 \times 9^1/_2$ inches. The same rules apply for filing reply comments. See the accompanying sidebar, "Hints on Filing Comments with the FCC."

All comments are reviewed by the FCC staff and play a role in the ultimate decision made by the FCC. Therefore, it's important to get them in by deadline. Comments not filed on time may not be considered by the FCC in its deliberations. Occasionally, a Notice of Inquiry (NOI) may be issued or combined with an NPRM. An NOI will set forth the FCC's concern over a particular matter and solicit comments and suggestions for the world at large as to whether adoption, amendment or repeal of a rule may be desirable. The same notice and comment procedures are followed with NOIs as with NPRMs.

Copies of FCC NPRMs concerning Amateur Radio are available from ARRL HQ for an SASE. Summaries or the word-for-word text of NPRMs are printed in the *Federal Register*, which is available in most local libraries. Listen to W1AW and watch *QST*, *The ARRL Letter* and other ham publications for information on the latest FCC happenings concerning Amateur Radio. Copies of FCC documents may be obtained, for a fee, from the FCC's contractor for records duplication: The

International Transcription Service, Suite 140, 2100 M St, NW, Washington, DC 20037, phone 202-857-3800. FCC NPRMs are usually posted on the FCC Internet sites; see Appendix 12.

The Report and Order

Once all the comments are reviewed (for amateur matters, this is usually done by the Private Wireless Division of the WTB), a draft Report and Order will be prepared by that branch and submitted to the Chief of the WTB. Then the document must work its way up the hierarchy, surviving checkpoints along the way. A draft Report and Order must be approved at each level within the bureau and then coordinated with other interested bureaus or offices within the FCC before formally being presented to the commissioners for their consideration and approval. A Report and Order is just what it sounds like—the Commission issues a detailed report of its findings and an order is issued based on the judgment of the commissioners (as recommended by the staff).

Specific rules changes are listed. Revisions of the Report and Order take place right up to the moment it is adopted by the FCC commissioners, usually after due consideration by the commissioners at an "Open Meeting," a regular meeting of the commissioners that has a published agenda and is open to the public. During these open meetings, FCC staff members most familiar with the subject at hand make the presentation. The commissioners may decide to adopt the Report and Order as presented to them by the staff, order the staff to make specific revisions in the document and then adopt it, or terminate the NPRM without amending any rule or taking any action.

The rule or rules finally adopted need not be identical to the original proposal. The courts have held that there must merely be some relationship between the proposal and the rule finally adopted, and that the FCC must have considered all the relevant comments that were submitted. The Administrative Procedure Act requires that there must be a concise general statement of the basis and purpose incorporated in the rules. Sometimes these "concise" statements of why the FCC did what it did are anything but, going on and on in excruciating detail. What the FCC is doing is putting all its cards on the table to show it made a rational decision—this is done to avoid problems with Congress (which can exert pressure on the FCC because Congress controls the FCC's appropriations, that is, the purse strings) and judicial review of its decision (see "Court Appeals," below). The FCC tends to be sensitive to Congress and the courts, so it carefully frames its Reports and Orders and the like to ensure that it has stated its case as persuasively as possible.

Generally, a Report and Order is the end of the line for a particular rules change. The rules become law under the Administrative Procedure Act 30 days after publication in the *Federal Register*.

Due Process

You might wonder why this process takes so long and if it's worth it. As has been said many times, this is the price we pay for democracy. We are saddled with these elaborate and often slow procedures because we have a free society. All interested persons have an opportunity to hear and to be heard on a given issue— it's called due process of law, which is embodied in the US Constitution. These tedious procedures that allow for public participation at the expense of speedy decision making are desirable in our system. It's one of the things that distinguishes us from the totalitarian regimes around the world, and makes the US the world leader in freedom.

Petitions for Reconsideration

§405 of the Communications Act, as amended, affords "a person aggrieved or whose interests are adversely affected" by an order (a new or amended rule, etc) the right to petition the FCC for reconsideration. The petitioner must state specifically why the FCC's action should be changed. A Petition for Reconsideration must be filed within 30 days of the date of the public notice of the final FCC action, which generally means within 30 days of its publication in the *Federal Register*.

A Petition for Reconsideration usually is referred to the same bureau that prepared the original Report and Order. It is virtually impossible to obtain a favorable action on a Petition for Reconsideration if no new facts are presented. Occasionally, however, if the petitioner shows good cause, the FCC may grant the Petition for Reconsideration and modify the earlier order.

Comments in opposition to a Petition for Reconsideration must be filed within 15 days of the date of public notice of the petition's filing, again usually the date of publication in the *Federal Register*. Replies to an opposition must be filed within 10 days after the time for filing oppositions has expired.

The format and number of copies required for reconsideration petitions are the same as described above, although they may not exceed 25 double-spaced typewritten pages. Oppositions to Petitions for Reconsideration are also limited to 25 double-spaced typewritten pages, while replies to an opposition are limited to 10 double-spaced typewritten pages.

The effective date of a new or amended rule is not automatically postponed by the filing of a Petition for Reconsideration. If a stay of the effective date of the rule or amendment is desired, the petitioner must specifically request it in a "Motion to Stay" and must show good cause (such as irreparable harm) why the rule should not go into effect.

Court Appeals

Judicial review of an FCC decision may be sought under §402(a) of the Communications Act and §§701-706 of the Administrative Procedure Act, as long as review is sought over a *final* decision. Appeals from decisions and orders of the Commission are, in cases listed in §402(b) of the Communications Act, to be taken to the United States Court of Appeals for the District of Columbia. The court may either affirm (approve) the action of the FCC, reverse it, or send it back to the FCC for further consideration.

Generally, the court will be quite deferential to an agency such as the FCC, because of the technical nature of the communications world, so the chances of the aggrieved person prevailing on appeal are remote. The standard the court uses is whether the agency action is "arbitrary, capricious or an abuse of discretion." If the court can find a reasonable or rational basis for an agency action (which is, again, why FCC Report and Orders try to address every conceivable issue), no abuse of discretion will be found. A court might set aside a rule making action if, for example, the FCC exceeded the authority delegated to it by Congress or under the Constitution, if it failed to properly follow its notice-and-comment procedures, on the record (that is, the agency's fact-finding was unreasonable), or in the event there was a prejudicial error.

As mentioned, the court generally follows a policy of not substituting its judgment for that of the agency because of the agency's presumed expertise in that field. Moreover, courts are much more comfortable with issues of law (procedure, applicability of statutes, and so on) and are reluctant to address the facts and details of a particular controversy. The standard for evaluating a question of fact would be whether the FCC's action was supported by substantial evidence on the record; in questions of law, the court would be much more inclined to substitute its own judgment for that of the FCC or other agency. As such, in a factual determination, the court will rarely go beyond the facts and evidence already put on the record during the agency rule making.

In the relatively rare event that the court remands a rule making matter back to the FCC for further consideration, it will usually be sent back to the same bureau that handled it originally.

It is clear from this discussion that Petitions for Reconsideration and appeals to the courts are not guaranteed to bring about changes in rules adopted in the rule making process. Often the most practical course is to file a new petition for rule making after experience has been gained with the new or amended rules and it can be demonstrated that the rule, as amended, is not working out as planned and therefore not in the best interests of the amateur community.

The FCC and You

This chapter talks about your right as a citizen of the United States to participate in the FCC's rule-making process. The ARRL doesn't want to undercut your right to participate in the process, but please exercise common sense and think before you leap. For example, say you've been ragchewing on the same 75-meter frequency at the same time with the same stations for 364 uninterrupted days. On day 365, you can't complete your contact because of QRM from a contest. You should *not* then run to your word processor and draft a petition to the FCC to prohibit contesting. That would be totally out of proportion to such a temporary situation.

A responsibility of Amateur Radio's tradition of self-regulation as discussed in this book means that at all times hams must minimize any tendency to complain or formally petition the FCC for narrow or restrictive rule interpretations, particularly for emotional reasons. Otherwise, unfavorable precedents may be set that will be harmful to the Amateur Radio Service. A call to the FCC may result in a hasty or uninformed legal opinion by an FCC staff member with no real-world operating experience. This opinion would be rendered in good faith, but you can expect that it would be formulated without consideration of all the necessary details. The FCC staff faces a formidable workload and is doing the best it can under the circumstances; the FCC wants hams to resolve operational issues by themselves, as provided for in Part 97.

In terms of a rule making already pending or in a highly important, significant situation where you as a radio amateur feel that it's crucial to submit a petition for rule making on your own, it is best to contact the League *first*. The ARRL acts as the clearinghouse for rule making matters and painstakingly works towards gauging a consensus on an issue *before* formally contacting the FCC. The League is a representative democracy; when the League determines that a consensus exists, it initiates proposals for the good of the amateur service, such as the ARRL petition for rule making that culminated in Novice Enhancement or the codeless Technician license. The League monitors *all* FCC proposals, whether they directly or indirectly affect Amateur Radio, publicizes them extensively in *QST*, *The ARRL Letter,* W1AW bulletins, and elsewhere, and submits comments on behalf of the amateur community on every one, as appropriate. It also encourages individual amateurs to file comments on important matters.

Before you participate in an FCC proceeding, check with the Regulatory Information Branch at ARRL HQ or your ARRL Director to find out more about that particular issue and its status. The ARRL knows how to best approach the FCC— eight decades of crusading for amateurs' rights have given the League unexcelled experience and a clear understanding of government regulatory processes. This is a major reason amateurs support the ARRL and have made it a respected voice in FCC deliberations.

Conclusion

New editions of the *FCC Rule Book* are published to keep pace with changes in the rules. In addition, the ARRL maintains electronic versions of Part 97 on its HIRAM BBS, Internet Info server, ftp and WWW sites, and various on-line services. These versions are updated with every rule change; contact the Regulatory Information Branch for details. The rules have evolved over time and will continue to do so. The rules are dynamic because Amateur Radio is dynamic—constantly changing to meet and create new methods and technologies to better serve individuals and society at large.

CHAPTER 16

Rules Compliance, Self-Regulation and You

A mong the many significant provisions of the Communications Amendments Act of 1982 [Public Law 97-259] is one that authorizes the FCC to use amateur volunteers to monitor the airwaves for rules violations. This is crucial in maintaining order and the traditional high standards on the amateur bands, as governmental belt-tightening deprives the FCC's Compliance and Information Bureau (CIB) of the necessary resources to monitor the amateur bands to any great extent. Moreover, the Volunteer Examiner Program and the creation of the codeless Technician license has helped the Amateur Radio Service to grow, creating an even greater potential need for monitoring of amateur band activity.

Here's what Congress said when finalizing the bill:

> The Amateur Radio Service has been praised for being self-regulated. The Commission has reported that less time has been devoted to monitoring and regulating the Amateur Service than to any other service because of its self-policing and discipline.

> One primary purpose of the Conference Substitute is to provide the Federal Communications Commission with the authority to implement various programs which will result in improvements in administration of the Amateur Radio Service and to cut the costs thereof. It will further allow the Amateur Radio Service to continue its tradition as the most self-regulated radio service in the United States, and to become to some extent self-administered, requiring even less expenditure to government time and effort than in the past.

The Amateur Auxiliary

When they implemented volunteer monitoring, the FCC's Compliance and Information Bureau (then known as the Field Operations Bureau, or FOB) recognized the value of the organized and disciplined amateur community through its membership organization, the ARRL. They recognized the long history and tradition of the ARRL's Official Observer (OO) program and the League's extensive Field Organization.

To make good use of this, the CIB created the Amateur Auxiliary. The CIB and the ARRL signed an agreement outlining the objectives and nature of the program and the commitment by the ARRL to administer it.

The Amateur Auxiliary addresses both *maintenance monitoring* and *amateur-to-amateur interference*. Maintenance monitoring is conducted through the enhanced OO program, while local amateur-to-amateur interference is handled by specifically authorized *Local Interference Committees*. These committees are sanctioned by the ARRL Section Manager for the geographical area concerned.

The Amateur Auxiliary is administered by the League's Section Managers and OO Coordinators with support from ARRL HQ.

General Objectives/Concept

The general objectives of the Amateur Auxiliary are to:
1. Foster a wider knowledge of, and better compliance with, the laws, rules and regulations governing the Amateur Radio Service;
2. Extend the concepts of self-regulation and self-administration of the Service;
3. Enhance the opportunity for individual amateurs to contribute to the public welfare as outlined in the basis and purpose of the Amateur Radio Service;
4. Enable the FCC's Compliance and Information Bureau to efficiently and effectively utilize its manpower and resources.

The role of the Amateur Auxiliary is to give technical and operational advice and other assistance to amateurs who are receptive. The task is not to find fault but to identify causes and effects, and to find ways to achieve solutions. *The mission is not direct enforcement of the rules.* Because the boundary between observation and enforcement is not always obvious, mature judgment is clearly required of the Auxiliary's members and leadership. The Auxiliary, to be viable and effective, must avoid the appearance of enforcement. It must also avoid the appearance of having a vested interest in any specific type of amateur operations or of being highly sympathetic to amateur groups that advocate specific activities or causes.

Training and Certification

Only those amateurs who have successfully demonstrated a thorough knowledge of, and skill in, volunteer monitoring are enrolled as members of the Amateur Auxiliary.

The appointment procedure is as follows:
1. All Official Observer and Local Interference Committee Chairman appointment inquiries are sent to the Section Manager (SM), since all OO appointments are initiated by the Section Manager or the SM's designee, the Official Observer Coordinator (OOC).
2. The SM or OOC sends an appointment card to ARRL HQ, indicating "RECOMMENDATION."
3. ARRL HQ sends the candidate the *Amateur Auxiliary Training Manual*, other supplementary literature, and an open-book examination paper. After the candidate reviews the material, he or she completes the exam and returns it to ARRL HQ for grading.
4. The OO is notified that he or she has successfully completed the training and exam, or is sent a re-test with subject areas needing further study pointed out. A copy of this notification goes to the Section Manager who then sends the successful candidate the official appointment certificate.

Compliance Level

The objective of the Amateur Auxiliary is to foster a satisfactory level of compliance in the Amateur Radio Service. To determine compliance trends, Amateur Auxiliary data is recorded and analyzed at ARRL HQ. This helps pinpoint areas in need of more or less attention and also enables the Auxiliary to report on the impact of the program to the FCC. This accountability enables the Auxiliary to continually sharpen skills and direct its efforts to those areas where they will do the most good.

Expectations

The Amateur Auxiliary cannot cure all the ills that affect the Amateur Service. Some of these ills are clear violations; others are not. The ability to differentiate is important. The general amateur public must not think volunteer monitoring can solve all these on-the-air problems (both real and perceived). For example, certain cases of malicious interference and harassment are prevalent on repeaters in some parts of the country. These problems cannot be cured swiftly or easily. Much evidence gathering is involved; this can be a time-consuming process.

Even more difficult for many amateurs to understand is how to deal with advocates of "free thinking": hams who don't comply with gentlemen's agreements, and so on. These individualists delight in creating chaos. Unfortunately, such types feed on the ignorance of others and thrive on baiting the listening audience into discussions of who is going to call the monitoring station first, who had the frequency first, and so on. Often the individualists are far better versed in the rules and regulations than those who confront them on the air. Too often this leads to frustration manifested by catcalls, whistles, carriers and general chaos on the frequency—exactly the desired effect. This continues to fan the flames, and can lead to the situation where the true violations are now being perpetrated by the hams in the audience. To expect the Auxiliary to solve such problems is not realistic.

Technical and Operating Discrepancies

The OO program is intended to note discrepancies and call them to the attention of those who very often are unaware of the discrepancy or who do not realize that what they are doing is potentially in violation. For example, it is quite common for an amateur to be unaware of a harmonic or spurious signal outside the amateur bands strong enough to be heard in distant places. Although new entry level licensees are especially prone to this discrepancy, higher-grade licensees, especially those running increased power, are far from immune to excessive harmonic radiation.

Often the operator of a station with key clicks and chirp on his CW signals is unaware they exist. Broad signals, with by-products of splatter and distortion, are frequent on the phone bands, while FM signals are prone to overdeviation. These are all technical difficulties that are watched for and are subject to notification. If you are the recipient of such a notification, please remember that the OO has sent it in a spirit of helpfulness, not fault finding. It is not an FCC notice, and no reply is necessary.

Official Observer Advisory Notice

The primary notification tool of the OO is the Official Observer Advisory Notice, Form FSD-213. This card is used to notify stations of operating discrepancies and usually includes a reference to the appropriate FCC regulation.

To avoid the impression that the notice arises from personal interest, Official Observers shouldn't send notices to someone interfering with them or a net or repeater they are operating on. This is not generally good for relations with the

OFFICIAL OBSERVER ADVISORY NOTICE

Radio: _____, your call was heard calling/working _____ at _____UTC.

Date: _____ 19 ___ Frequency: _____kHz. Mode _____. Your RST ____.

The following is noted in the interest of maintaining Amateur Radio's reputation for good operating/technical practices: 1❑FREQUENCY INSTABILITY 2❑CHIRP 3❑SPURIOUS 4❑HARMONIC 5❑HUM 6❑KEY CLICKS 7❑BROAD SIGNAL 8❑DISTORTED AUDIO 9❑OVER DEVIATION 10❑OUT OF BAND 11❑IMPROPER ID 12❑LANGUAGE 13❑CAUSING INTERFERENCE 15❑CARRIER 20❑OTHER

Remarks: _____

_____

_____

Please refer to FCC Regulation _____. Please take a few minutes to determine what equipment factors or operating practices might have contributed to this apparent departure from the rule or the good amateur practice standard. The intent of this notice is to alert you to the above noted operating condition. NO REPLY IS NECESSARY. The undersigned ARRL Official Observer has fulfilled this helping role by simply alerting you, and is not *required* to reply to any correspondence. Thank you for your attention and any cooperative efforts to enhance the high standards of the Amateur Radio Service which we all share with pride.

FSD-213(695)   Signature _____   Call _____

---

amateurs the program is trying to serve. On the other hand, there is no harm with an OO asking another Observer to take an independent look if a problem persists.

OOs should project a friendly spirit on their advisory notices. The objective is to bring about compliance with the regulations by friendly persuasion.

Good Operator Report

Official Observers may also report on stations heard that have outstandingly *good* signals or operating procedures. While in today's world of high-quality commercial radio equipment the radio owner is less likely to be responsible for the technical excellence of his signal, operating skill is still a matter of individual cultivation. These reports should not be sent out wholesale, but reserved for those amateurs who set an example

---

Good Operator Report ■

Radio: _____, your call was heard calling/working _____ at _____ UTC.

Date: _____199 _____Frequency _____ kHz. Mode _____. Your RST _____.

We thought you would like to know . . .

That this Official Observer has noted your EXCELLENT radio signal quality/operating procedure as a fine example for all radio amateurs.

Remarks: _____

_____

_____

This observation by the undersigned ARRL Official Observer is a function of the Amateur Auxiliary to the FCC's Compliance and Information Bureau. This Observer thanks you for your excellent example of good amateur practice for others in the Amateur Radio Service. Keep up the good work.

FSD-15(3/95)   Signature _____ Call _____

for the rest of the community by displaying the *best* that Amateur Radio has to offer.

Correspondence

Occasionally an OO will become involved in a "pen pal" exchange as a result of a notification card. Observers are not obligated to solve all of a recipient's problems, but many enjoy the opportunity to be helpful. Some observers go to considerable trouble to provide assistance. There are, however, practical limitations to an OO's involvement in the problems of an individual amateur. Follow-up correspondence may refer to the services of the ARRL Field Technical Information Service. These volunteers can suggest corrective technical action.

Difficult Cases

The Official Observer is usually involved with maintenance monitoring, i.e. scanning the bands to assist in *maintaining* the high reputation of self-regulation that the Amateur Service has earned. Occasionally, the OO will encounter the somewhat more controversial or unprofessional side of amateur operations, such as obscenity, indecency, false signals or willful or malicious interference. A great deal of tact must be used by OOs when sending notices for these discrepancies. In fact, the wisdom of sending notices for isolated instances is open to question at times.

If such a case develops into a prolonged case of amateur-to-amateur interference or other substantive violation, then it can become a case beyond the scope of the individual OO. If this is an HF case, such as interference caused to or by nets, then the Official Observer would almost certainly refer such a case to a higher level of the Amateur Auxiliary. Consultation with the section OO Coordinator is in order and, at the discretion of the Coordinator, the ARRL Headquarters staff may be involved.

Under no circumstances should the OO initiate direct contact with an FCC monitoring facility. Decision to make such contact is made at a higher level. Only the most extreme cases will be brought to the attention of the Compliance and Information Bureau, and then only under the professional guidance of FCC personnel in strict accordance with the established referral policy of the Auxiliary.

Petty cases of interference should be solved by amateurs and not referred to the FCC. Historically, FCC prosecution of just a few hard-core cases has had a marked positive effect on the behavior of those who might deliberately abuse our privileges. Short of FCC involvement, the OOC can advise the offender with a "heavy-duty" advisory notice; this should be used by the OOC before any FCC referral is made.

Local Interference Committees

Interference problems generated on VHF or UHF repeaters are primarily local problems requiring local resolution. The Amateur Auxiliary mechanism for dealing with any local amateur-to-amateur interference is the Local Interference Committee (LIC).

The LIC is within the purview of the Section Manager's overall Amateur Auxiliary program. The LIC gains official standing through the Section Manager. It is, in effect, a "group appointment" of the Section Manager, with specific authorization to deal with local amateur-to-amateur interference.

The prospective Committee makes a brief written proposal to the Section

Manager outlining the nature of the Committee structure. This proposal should include the following key elements:

1. Statement of purpose and reference to participation in the Amateur Auxiliary.
2. Area of jurisdiction or specific task, including a statement of impartiality.
3. Any appropriate operating instructions.
4. How long the committee plans to operate.
5. Name of the members of the committee, with a designated chairman.
6. Space for the committee chairman's signature.
7. Space for Section Manager's signature of authorization.

The Make-Up of the Committee

LIC members should be amateurs with experience in the installation and operation of repeater systems. Amateurs with VHF and/or UHF direction-finding equipment and experience are also very helpful. Members of the committee should preferably be selected by an area council of radio clubs, if such an organization exists, or by the clubs particularly interested in VHF, UHF and repeater operations if there is no council. The organization of the committee would be determined by the council, clubs, or committee members. The prime criterion for membership is that the members of the committee must be respected and accepted by the general amateur community in the area.

Whenever possible, at least one attorney, preferably an Amateur Radio operator, experienced with criminal trial practice and rules of evidence, should be a member of or available to the committee.

General LIC Procedures

Complaints of deliberate interference are received and investigated by the committee. The investigation might include preparation of audio-tape recordings of the interfering or objectionable transmissions, oscilloscope photos and other recordings showing the switching characteristics of the carrier, and results of direction-finding investigations. Wherever possible, all material should be sufficiently detailed and complete.

If such committees are to be effective, their investigations, as well as those of the FCC, must be *confidential*. The committees must be willing and able to *resist demands* for action from amateurs who are not familiar with what is being done.

Evidence Gathering

When all avenues of approach to a difficult, serious case have been exhausted within Auxiliary channels, the matter may be brought to the attention of the ARRL Headquarters staff. If it is deemed appropriate by the Headquarters staff to send the case to the FCC, ARRL HQ will work with the Auxiliary volunteers in properly constructing a package of evidence. OOs must conduct monitoring activities and assemble information in such a way that it is (1) useful evidence in cases of repeated or intentional rule violations, in which it may be used later in license suspension or revocation hearings, and (2) available should the OO have to protect himself against a defamation (libel or slander) action. Although both of these situations will be rare, it is difficult to predict, in any given situation, what the outcome of an Auxiliary report will be. OOs should prepare each OO report as though they will be called on to testify at an FCC hearing as to actions which gave rise to the enforcement proceedings.

Prior to the enactment of the Communications Amendments Act of 1982, amateurs were unable to provide voluntary monitoring services to the government, or to disclose to third parties what was heard on the air. Now, those limitations have been removed, and the evidence gathered by volunteers can be relied on by the FCC and used directly in enforcement proceedings. It is no longer necessary for the FCC staff to duplicate the monitoring and DFing done by volunteers.

To make sure the evidence is useful, and at the same time protect against a lawsuit based on an accusation that defamatory remarks were made about another amateur, OOs must first restrict all monitoring activities to fit squarely within the statutory authority granted to the FCC by the Communications Amendments Act of 1982. These are as follows:

(I) The detection of improper Amateur Radio transmissions;

(II) The conveyance to Commission personnel of information which is essential to the enforcement of the Communications Act or regulations prescribed by the FCC, relating to the Amateur Radio Service; and

(III) The issuance of advisory notices to persons who apparently have violated any provision of the Act or regulations prescribed by the FCC relating to the Amateur Radio Service.

Any activities not aimed specifically at one of these purposes exceeds the statutory authority of the monitor and may result in having evidence gathered under such circumstances ruled inadmissible as having been provided by an "incompetent" witness of the matter asserted.

Conclusion

The bottom line is that the Auxiliary can lead a horse to water, and with a nod of thanks, the horse usually drinks as it is in his own best interest. The Auxiliary can't *make* him drink, but if the situation is serious enough, the FCC can and will.

While there's always room for improvement, the Amateur Auxiliary continues to work well in helping the amateur community keep its own shop in order. With limited government involvement, it's about the only game in town, and is deserving of every amateur's support. For more information on the Amateur Auxiliary program, contact your ARRL Section Manager (listed on page 8 of every *QST*), or ARRL Headquarters.

PART 97
Amateur Radio Service

Here is the complete text of Part 97 of the FCC regulations, the rules that govern the Amateur Service. One problem the FCC has is making sure all applicable references are updated when the rules are changed. Occasionally they miss some, as when they created the Technician Plus license class. With eight pages of changes they managed to get most citations, but forgot to add the words "Technician Plus class" to 97.201(a), 97.203(a), and 97.205(a) to make it clear that Technician Plus licensees may also establish and/or be control operators of repeaters, beacons, and auxiliary stations. Similarly, in 97.307(f) (9) and (10), "Technician" should be "Technician Plus." The interim ID rules in 97.119(e) don't give a separate indicator for Novices or Technicians upgrading to Technician Plus.

In 1995 the FCC changed the name of the Automatic Vehicle Monitoring Systems (AVMS) to the Location and Monitoring Service (LMS) in Part 90. This change was not made in 97.303(g)(l), which mentions the AVMS. Lastly, in 97.303(k), it should be 3.525 GHz instead of 3525 GHz.

For those who have access to the Internet, the ARRL Regulatory Information Branch maintains current versions of Part 97 on the following sites—World Wide Web (http://www.arrl.org/); FTP (oak.oakland.edu), ARRL's automated information server (info@arrl.org), and our landline BBS (860-594-0306). RIB also maintains current versions on CompuServe's Hamnet Forum, and the Ham Radio section of America OnLine. These versions are updated whenever a rule change takes effect.

(current as of April 11, 1996)

Subpart A—General Provisions

§97.1 Basis and purpose.

The rules and regulations in this Part are designed to provide an amateur radio service having a fundamental purpose as expressed in the following principles:

(a) Recognition and enhancement of the value of the amateur service to the public as a voluntary noncommercial communication service, particularly with respect to providing emergency communications.

(b) Continuation and extension of the amateur's proven ability to contribute to the advancement of the radio art.

(c) Encouragement and improvement of the amateur service through rules which provide for advancing skills in both the communications and technical phases of the art.

(d) Expansion of the existing reservoir within the amateur radio service of trained operators, technicians, and electronics experts.

(e) Continuation and extension of the amateur's unique ability to enhance international goodwill.

§97.3   Definitions.

(a) The definitions of terms used in Part 97 are:

(1) *Amateur operator.* A person holding a written authorization to be the control operator of an amateur station.

(2) *Amateur radio services.* The amateur service, the amateur-satellite service and the radio amateur civil emergency service.

(3) *Amateur-satellite service.* A radiocommunication service using stations on Earth satellites for the same purpose as those of the amateur service.

(4) *Amateur service.* A radiocommunication service for the purpose of self-training, intercommunication and technical investigations carried out by amateurs, that is, duly authorized persons interested in radio technique solely with a personal aim and without pecuniary interest.

(5) *Amateur station.* A station in an amateur radio service consisting of the apparatus necessary for carrying on radiocommunications.

(6) *Automatic control.* The use of devices and procedures for control of a station when it is transmitting so that compliance with the FCC Rules is achieved without the control operator being present at a control point.

(7) *Auxiliary station.* An amateur station, other than in a message forwarding system, that is transmitting communications point-to-point within a system of cooperating amateur stations.

(8) *Bandwidth.* The width of a frequency band outside of which the mean power of the transmitted signal is attenuated at least 26 dB below the mean power of the transmitted signal within the band.

(9) *Beacon.* An amateur station transmitting communications for the purposes of observation of propagation and reception or other related experimental activities.

(10) *Broadcasting.* Transmissions intended for reception by the general public, either direct or relayed.

(11) *Call sign system.* The method used to select a call sign for amateur station over-the-air identification purposes. The call sign systems are:

(i) Sequential call sign system. The call sign is selected by the FCC from an alphabetized list corresponding to the geographic region of the licensee's mailing address and operator class. The call sign is shown on the license. The FCC will issue public announcements detailing the procedures of the sequential call sign system.

(ii) Vanity call sign system. The call sign is selected by the FCC from a list of call signs requested by the licensee. The call sign is shown on the license. The FCC will issue public announcements detailing the procedures of the vanity call sign system.

(12) *Control operator.* An amateur operator designated by the licensee of a station to be responsible for the transmissions from that station to assure compliance with the FCC Rules.

(13) *Control point.* The location at which the control operator function is performed.

(14) *CSCE.* Certificate of successful completion of an examination.

(15) *Earth station.* An amateur station located on, or within 50 km of the Earth's surface intended for communications with space stations or with other Earth stations by means of one or more other objects in space.

(16) *EIC.* Engineer in Charge of an FCC Field Facility.

(17) *External RF Power Amplifier.* A device capable of increasing power output when used in conjunction with, but not an integral part of, a transmitter.

(18) *External RF power amplifier kit.* A number of electronic parts, which, when assembled, is an external RF power amplifier, even if additional parts are required to complete assembly.

(19) *FAA.* Federal Aviation Administration.

(20) *FCC.* Federal Communications Commission.

(21) *Frequency coordinator.* An entity, recognized in a local or regional area by amateur operators whose stations are eligible to be auxiliary or repeater stations, that recommends transmit/receive channels and associated operating and technical parameters for such stations in order to avoid or minimize potential interference.

(22) *Harmful interference.* Interference which endangers the functioning of a radionavigation service or of other safety services or seriously degrades, obstructs or repeatedly interrupts a radiocommunication service operating in accordance with the Radio Regulations.

(23) *Indicator.* Words, letters or numerals appended to and separated from the call sign during the station identification.

(24) *Information bulletin.* A message directed only to amateur operators consisting solely of subject matter of direct interest to the amateur service.

(25) *International Morse code.* A dot-dash code as defined in International Telegraph and Telephone Consultative Committee (CCITT) Recommendation F.1 (1984), Division B, I. Morse Code.

(26) *ITU.* International Telecommunication Union.

(27) *Line A.* Begins at Aberdeen, WA, running by great circle arc to the intersection of 48° N, 120° W, thence along parallel 48° N, to the intersection of 95° W, thence by great circle arc through the southernmost

point of Duluth, MN, thence by great circle arc to 45° N, 85° W, thence southward along meridian 85° W, to its intersection with parallel 41° N, thence along parallel 41° N, to its intersection with meridian 82° W, thence by great circle arc through the southernmost point of Bangor, ME, thence by great circle arc through the southernmost point of Searsport, ME, at which point it terminates.

(28) *Local control.* The use of a control operator who directly manipulates the operating adjustments in the station to achieve compliance with the FCC Rules.

(29) *Message forwarding system.* A group of amateur stations participating in a voluntary, cooperative, interactive arrangement where communications are sent from the control operator of an originating station to the control operator of one or more destination stations by one or more forwarding stations.

(30) *National Radio Quiet Zone.* The area in Maryland, Virginia and West Virginia bounded by 39° 15' N on the north, 78° 30' W on the east, 37° 30' N on the south and 80° 30' W on the west.

(31) *Physician.* For the purposes of this Part, a person who is licensed to practice in a place where the amateur service is regulated by the FCC, as either a Doctor of Medicine (MD) or a Doctor of Osteopathy (DO).

(32) *Question pool.* All current examination questions for a designated written examination element.

(33) *Question set.* A series of examination questions on a given examination selected from the question pool.

(34) *Radio Regulations.* The latest ITU *Radio Regulations* to which the United States is a party.

(35) *RACES (radio amateur civil emergency service).* A radio service using amateur stations for civil defense communications during periods of local, regional or national civil emergencies.

(36) *Remote control.* The use of a control operator who indirectly manipulates the operating adjustments in the station through a control link to achieve compliance with the FCC Rules.

(37) *Repeater.* An amateur station that simultaneously retransmits the transmission of another amateur station on a different channel or channels.

(38) *Space station.* An amateur station located more than 50 km above the Earth's surface.

(39) *Space telemetry.* A one-way transmission from a space station of measurements made from the measuring instruments in a spacecraft, including those relating to the functioning of the spacecraft.

(40) *Spurious emission.* An emission, on frequencies outside the necessary bandwidth of a transmission, the level of which may be reduced without affecting the information being transmitted.

(41) *Telecommand.* A one-way transmission to initiate, modify, or terminate functions of a device at a distance.

(42) *Telecommand station.* An amateur station that transmits communications to initiate, modify, or terminate functions of a space station.

(43) *Telemetry.* A one-way transmission of measurements at a distance from the measuring instrument.

(44) *Third-party communications.* A message from the control operator (first party) of an amateur station to another amateur station control operator (second party) on behalf of another person (third party).

(45) *VE.* Volunteer examiner.

(46) *VEC.* Volunteer-examiner coordinator.

(b) The definitions of technical symbols used in this Part are:

(1) *EHF* (extremely high frequency). The frequency range 30-300 GHz.

(2) *HF* (high frequency). The frequency range 3-30 MHz.

(3) *Hz.* Hertz.

(4) *m.* Meters.

(5) *MF* (medium frequency). The frequency range 300-3000 kHz.

(6) *PEP* (peak envelope power). The average power supplied to the antenna transmission line by a transmitter during one RF cycle at the crest of the modulation envelope taken under normal operating conditions.

(7) *RF.* Radio frequency.

(8) *SHF* (super-high frequency). The frequency range 3-30 GHz.

(9) *UHF* (ultra-high frequency). The frequency range 300-3000 MHz.

(10) *VHF* (very-high frequency). The frequency range 30-300 MHz.

(11) *W.* Watts.

(c) The following terms are used in this Part to indicate emission types. Refer to §2.201 of the FCC Rules, *Emission, modulation and transmission characteristics*, for information on emission type designators.

(1) *CW.* International Morse code telegraphy emissions having designators with A, C, H, J or R as the first symbol; 1 as the second symbol; A or B as the third symbol; and emissions J2A and J2B.

(2) *Data.* Telemetry, telecommand and computer communications emissions having designators with A, C, D, F, G, H, J or R as the first symbol; 1 as the second symbol; D as the third symbol; and emission J2D. Only a digital code of a type specifically authorized in this Part may be transmitted.

(3) *Image.* Facsimile and television emissions having designators with A, C, D, F, G, H, J or R as the first symbol; 1, 2 or 3 as the second symbol; C or F as the third symbol; and emissions having B as the first symbol; 7, 8 or 9 as the second symbol; W as the third symbol.

(4) *MCW.* Tone-modulated international Morse code telegraphy emissions having designators with A, C, D, F, G, H or R as the first symbol; 2 as the second symbol; A or B as the third symbol.

(5) *Phone.* Speech and other sound emissions having designators with A, C, D, F, G, H, J or R as the first symbol; 1, 2 or 3 as the second symbol; E as the third symbol. Also speech emissions having B as the first symbol; 7, 8 or 9 as the second symbol; E as the third symbol. MCW for the purpose of performing the station identification procedure, or for providing telegraphy practice interspersed with speech. Incidental tones for the purpose of selective calling or alerting or to control the level of a demodulated signal may also be considered phone.

(6) *Pulse.* Emissions having designators with K, L, M, P, Q, V or W as the first symbol; 0, 1, 2, 3, 7, 8, 9 or X as the second symbol; A, B, C, D, E, F, N, W or X as the third symbol.

(7) *RTTY.* Narrow-band direct-printing telegraphy emissions having des-

ignators with A, C, D, F, G, H, J or R as the first symbol; 1 as the second symbol; B as the third symbol; and emission J2B. Only a digital code of a type specifically authorized in this Part may be transmitted.

(8) *SS*. Spread-spectrum emissions using bandwidth-expansion modulation emissions having designators with A, C, D, F, G, H, J or R as the first symbol; X as the second symbol; X as the third symbol. Only a SS emission of a type specifically authorized in this Part may be transmitted.

(9) *Test*. Emissions containing no information having the designators with N as the third symbol. Test does not include pulse emissions with no information or modulation unless pulse emissions are also authorized in the frequency band.

§97.5 Station license required.

(a) The person having physical control of the station apparatus must have been granted a station license of the type listed in paragraph (b), or hold an unexpired document of the type listed in paragraph (c), before the station may transmit on any amateur service frequency from any place that is:

(1) Within 50 km of the Earth's surface and at a place where the amateur service is regulated by the FCC;

(2) Within 50 km of the Earth's surface and aboard any vessel or craft that is documented or registered in the United States; or

(3) More than 50 km above the Earth's surface aboard any craft that is documented or registered in the United States.

(b) The types of station licenses are:

(1) An operator/primary station license. One, but only one, operator/primary station license is granted to each person who is qualified to be an amateur operator. The primary station license is granted together with the amateur operator license. Except for a representative of a foreign government, any person who qualifies by examination is eligible to apply for an operator/primary station license. The operator/primary station license document is printed on FCC Form 660.

(2) A club station license. A club station license is granted only to the person who is the license trustee designated by an officer of the club. The trustee must be a person who has been granted an Amateur Extra, Advanced, General, Technician Plus, or Technician operator license. The club must be composed of at least two persons and must have a name, a document of organization, management, and a primary purpose devoted to amateur service activities consistent with this Part. The club station license document is printed on FCC Form 660.

(3) A military recreation station license. A military recreation station license is granted only to the person who is the license custodian designated by the official in charge of the United States military recreational premises where the station is situated. The person must not be a representative of a foreign government. The person need not have been granted an amateur operator license. The military recreation station license document is printed on FCC Form 660.

(4) A RACES station license. A RACES station license is granted only to the person who is the license custodian designated by the official re-

sponsible for the governmental agency served by that civil defense organization. The custodian must be the civil defense official responsible for coordination of all civil defense activities in the area concerned. The custodian must not be a representative of a foreign government. The custodian need not have been granted an amateur operator license. The RACES station license document is printed on FCC Form 660.

(c) The types of documents are:

(1) A *reciprocal permit for alien amateur licensee* (FCC Form 610-AL) issued to the person by the FCC.

(2) An amateur service license issued to the person by the Government of Canada. The person must be a Canadian citizen.

(d) A person who has been granted a station license of the type listed in paragraph (b), or who holds an unexpired document of the type listed in paragraph (c), is authorized to use in accordance with the FCC Rules all transmitting apparatus under the physical control of the station licensee at points where the amateur service is regulated by the FCC.

§97.7 Control operator required.

When transmitting, each amateur station must have a control operator. The control operator must be a person who has been granted an amateur operator/ primary station license, or who holds an unexpired document of the following types:

(a) A *reciprocal permit for alien amateur licensee* (FCC Form 610-AL) issued to the person by the FCC, or

(b) An amateur service license issued to the person by the Government of Canada. The person must be a Canadian citizen.

§97.9 Operator license.

(a) The classes of amateur operator licenses are: Novice, Technician, Technician Plus (until such licenses expire, a Technician Class license granted before February 14, 1991, is considered a Technician Plus Class license), General, Advanced, and Amateur Extra. A person who has been granted an operator license is authorized to be the control operator of an amateur station with the privileges of the operator class specified on the license.

(b) A person who has been granted an operator license of Novice, Technician, Technician Plus, General, or Advanced class and who has properly submitted to the administering VEs an application document, FCC Form 610, for an operator license of a higher class, and who holds a CSCE indicating that the person has completed the necessary examinations within the previous 365 days, is authorized to exercise the rights and privileges of the higher operator class until final disposition of the application or until 365 days following the passing of the examination, whichever comes first.

§97.11 Stations aboard ships or aircraft.

(a) The installation and operation of an amateur station on a ship or aircraft must be approved by the master of the ship or pilot in command of the aircraft.

(b) The station must be separate from and independent of all other radio apparatus installed on the ship or aircraft, except a common antenna may be shared with a voluntary ship radio installation. The station's transmissions must not cause interference to any other apparatus installed on the ship or aircraft.

(c) The station must not constitute a hazard to the safety of life or property. For a station aboard an aircraft, the apparatus shall not be operated while the aircraft is operating under Instrument Flight Rules, as defined by the FAA, unless the station has been found to comply with all applicable FAA Rules.

§97.13 Restrictions on station locations.

(a) Before placing an amateur station on land of environmental importance or that is significant in American history, architecture or culture, the licensee may be required to take certain actions prescribed by §1.1301 - 1.1319 of the FCC Rules.

(b) A station within 1600 m (1 mile) of an FCC monitoring facility must protect that facility from harmful interference. Failure to do so could result in imposition of operating restrictions upon the amateur station by an EIC pursuant to §97.121 of this Part. Geographical coordinates of the facilities that require protection are listed in § 0.121(c) of the FCC Rules.

§97.15 Station antenna structures.

(a) Unless the amateur station licensee has received prior approval from the FCC, no antenna structure, including the radiating elements, tower, supports and all appurtenances, may be higher than 61 m (200 feet) above ground level at its site.

(b) Unless the amateur station licensee has received prior approval from the FCC, no antenna structure, at an airport or heliport that is available for public use and is listed in the *Airport Directory* of the current *Airman's Information Manual* or in either the *Alaska* or *Pacific Airman's Guide and Chart Supplement*; or at an airport or heliport under construction that is the subject of a notice or proposal on file with the FAA, and except for military airports, it is clearly indicated that the airport will be available for public use; or at an airport or heliport that is operated by the armed forces of the United States; or at a place near any of these airports or heliports, may be higher than:

(1) 1 m above the airport elevation for each 100 m from the nearest runway longer than 1 km within 6.1 km of the antenna structure.

(2) 2 m above the airport elevation for each 100 m from the nearest runway shorter than 1 km within 3.1 km of the antenna structure.

(3) 4 m above the airport elevation for each 100 m from the nearest landing pad within 1.5 km of the antenna structure.

(c) An amateur station antenna structure no higher than 6.1 m (20 feet) above ground level at its site or no higher than 6.1 m above any natural object or existing manmade structure, other than an antenna structure, is exempt from the requirements of paragraphs (a) and (b) of this Section.

(d) Further details as to whether an aeronautical study is required or if the structure must be registered, painted, or lighted are contained in Part 17 of the FCC Rules, Construction, Marking, and Lighting of Antenna Structures. To request approval to place an antenna structure higher than the limits specified in paragraghs (a), (b), and (c) of this section, the licensee must notify the FAA using FAA Form 7460-1 and the structure owner must register the structure using FCC Form 854.

(e) Except as otherwise provided herein, a station antenna structure may be erected at heights and dimensions sufficient to accommodate amateur service communications. [State and local regulation of a station antenna structure must not preclude amateur service communications. Rather, it must reasonably accommodate such communications and must constitute the minimum practicable regulation to accomplish the state or local authority's legitimate purpose. See PRB-1, 101 FCC 2d 952 (1985) for details.]

§97.17 Application for new license or *reciprocal permit for alien amateur licensee.*

(a) Any qualified person is eligible to apply for an amateur service license.

(b) Each application for a new amateur service license must be made on the proper document:
   (1) FCC Form 610 for a new operator/primary station license;
   (2) FCC Form 610-A for a *reciprocal permit for alien amateur licensee*; and
   (3) FCC Form 610-B for a new amateur service club or military recreation station license.

(c) Each application for a new operator/primary station license must be submitted to the VEs administering the qualifying examination.

(d) Any eligible person may apply for a *reciprocal permit for alien amateur licensee.* The application document, FCC Form 610-A, must be submitted to the FCC, 1270 Fairfield Road, Gettysburg, PA 17325-7245.
   (1) The person must be a citizen of a country with which the United States has arrangements to grant reciprocal operating permits to visiting alien amateur operators is eligible to apply for a *reciprocal permit for alien amateur licensee.*
   (2) The person must be a citizen of the same country that issued the amateur service license.
   (3) No person who is a citizen of the United States, regardless of any other citizenship also held, is eligible for a *reciprocal permit for alien amateur licensee.*
   (4) No person who has been granted an amateur operator license is eligible for a *reciprocal permit for alien amateur licensee.*

(e) No person shall obtain or attempt to obtain, or assist another person to obtain or attempt to obtain, an amateur service license or *reciprocal permit for alien amateur licensee* by fraudulent means.

(f) One unique call sign will be shown on the license of each new primary, club, and military recreation station. The call sign will be selected by the sequential call sign system.

(g) Each application for a new club or military recreation station license must be submitted to the FCC, 1270 Fairfield Road, Gettysburg, PA 17325-7245. No new license for a RACES station will be issued.

§97.19 Application for a vanity call sign.

(a) A person who has been granted an operator/primary station license or a license trustee who has been granted a club station license is eligible to make application for modification of the license, or the renewal thereof, to show a call sign selected by the vanity call sign system. RACES and military recreation stations are not eligible for a vanity call sign.

(b) Each application for a modification of an operator/primary or club station license, or the renewal thereof, to show a call sign selected by the vanity call sign system must be made on FCC Form 610-V. The form must be submitted with the proper fee to the address specified in the Wireless Telecommunications Bureau Fee Filing Guide.

(c) Only unassigned call signs that are available to the sequential call sign system are available to the vanity call sign system with the following exceptions:

(1) A call sign shown on an expired license is not available to the vanity call sign system for 2 years following the expiration of the license.

(2) A call sign shown on a surrendered, revoked, set aside, cancelled, or voided license is not available to the vanity call sign system for 2 years following the date such action is taken.

(3) Except for an applicant who is the spouse, child, grandchild, stepchild, parent, grandparent, stepparent, brother, sister, stepbrother, stepsister, aunt, uncle, niece, nephew, or in-law, and except for an applicant who is a club station license trustee acting with the written consent of at least one relative, as listed above, of a person now deceased, the call sign shown on the license of a person now deceased is not available to the vanity call sign system for 2 years following the person's death, or for 2 years following the expiration of the license, whichever is sooner.

(d) The vanity call sign requested by an applicant must be selected from the group of call signs corresponding to the same or lower class of operator license held by the applicant as designated in the sequential call sign system.

(1) The applicant must request that the call sign shown on the current license be vacated and provide a list of up to 25 call signs in order of preference.

(2) The first assignable call sign from the applicant's list will be shown on the license grant. When none of those call signs are assignable, the call sign vacated by the applicant will be shown on the license grant.

(3) Vanity call signs will be selected from those call signs assignable at the time the application is processed by the FCC.

(4) A call sign designated under the sequential call sign system for Alaska, Hawaii, Caribbean Insular Areas, and Pacific Insular areas will be assigned only to a primary or club station whose licensee's mailing address is in the corresponding state, commonwealth, or island. This limitation does not apply to an applicant for the call sign as the spouse, child, grandchild, stepchild, parent, grandparent, stepparent, brother, sister, stepbrother, stepsister, aunt, uncle, niece, nephew, or in-law, of the former holder now deceased.

§97.21 Application for a modified or renewed license.

(a) A person who has been granted an amateur station license that has not expired:

(1) Must apply for a modification of the license as necessary to show the correct mailing address, licensee name, club name, license trustee name, or license custodian name. The application document must be submit-

ted to: FCC, 1270 Fairfield Road, Gettysburg, PA 17325-7245. For an operator/primary station license, the application must be made on FCC Form 610. For a club, military recreation, or RACES station license, the application must be made on FCC Form 610-B.

(2) May apply for a modification of the license to show a higher operator class. The application must be made on FCC Form 610 and must be submitted to the VEs administering the qualifying examination.

(3) May apply for renewal of the license for another term. (The FCC may mail to the licensee an FCC Form 610-R that may be used for this purpose.)

 (i) When the license does not show a call sign selected by the vanity call sign system, the application may be made on FCC Form 610-R if it is received from the FCC. If the Form 610-R is not received from the FCC within 30 days of the expiration date of the license for an operator/primary station license, the application may be made on FCC Form 610. For a club, military recreation, or RACES station license, the application may be made on FCC Form 610-B. The application may be submitted no more than 90 days before its expiration to: FCC, 1270 Fairfield Road, Gettysburg, PA 17325-7245. When the application for renewal of the license has been received by the FCC at 1270 Fairfield Road, Gettysburg, PA 17325-7245 prior to the license expiration date, the license operating authority is continued until the final disposition of the application.

 (ii) When the license shows a call sign selected by the vanity call sign system, the application must be filed as specified in Section 97.19(b). When the application has been received at the proper address specified in the Wireless Telecommunications Bureau Fee Filing Guide prior to the license expiration date, the license operating authority is continued until final disposition of the application.

(4) May apply for a modification of the license to show a different call sign selected by the sequential call sign system. The application document must be submitted to: FCC, 1270 Fairfield Road, Gettysburg, PA 17325-7245. The application must be made on FCC Form 610. This modification is not available to club, military recreation, or RACES stations.

(b) A person who had been granted an amateur station license, but the license has expired, may apply for renewal of the license for another term during a 2 year filing grace period. The application document must be received by the FCC at 1270 Fairfield Road, Gettysburg, PA 17325-7245 prior to the end of the grace period. For an operator/primary station license, the application must be made on FCC Form 610. For a club, military recreation, or RACES station license, the application must be made on FCC Form 610-B. Unless and until the license is renewed, no privileges in the Part are conferred.

(c) Each application for a modified or renewed amateur service license must be accompanied by a photocopy (or the original) of the license document unless an application for renewal using FCC Form 610-R is being made, or unless the original document has been lost, mutilated or destroyed.

(d) Unless the holder of a station license requests a change in call sign, the same call sign will be assigned to the station upon renewal or modification of a station license.

(e) *A reciprocal permit for alien amateur licensee* cannot be renewed. A new *reciprocal permit for alien amateur licensee* may be issued upon proper application.

§97.23 Mailing address.

(a) Each application for a license and each application for a *reciprocal permit for alien amateur licensee* must show a mailing address in an area where the amateur service is regulated by the FCC and where the licensee or permittee can receive mail delivery by the United States Postal Service. Each application for a *reciprocal permit for alien amateur licensee* must also show the permittee's mailing address in the country of citizenship.

(b) When there is a change in the mailing address for a person who has been granted an amateur operator/primary station license, the person must file a timely application for a modification of the license. Revocation of the station license or suspension of the operator license may result when correspondence from the FCC is returned as undeliverable because the person failed to provide the correct mailing address.

(c) When a person who has been granted a *reciprocal permit for alien amateur licensee* changes the mailing address where he or she can receive mail delivery by the United States Postal Service, the person must file an application for a new permit. Cancellation of the *reciprocal permit for alien amateur licensee* may result when correspondence from the FCC is returned as undeliverable because the permittee failed to provide the correct mailing address.

§97.25 License term.

(a) An amateur service license is normally granted for a 10-year term.

(b) A *reciprocal permit for alien amateur licensee* is normally granted for a 1-year term.

§97.27 FCC modification of station license.

(a) The FCC may modify a station license, either for a limited time or for the duration of the term thereof, if it determines:

(1) That such action will promote the public interest, convenience, and necessity; or

(2) That such action will promote fuller compliance with the provisions of the Communications Act of 1934, as amended, or of any treaty ratified by the United States.

(b) When the FCC makes such a determination, it will issue an order of modification. The order will not become final until the licensee is notified in writing of the proposed action and the grounds and reasons therefor. The licensee will be given reasonable opportunity of no less than 30 days to protest the modification; except that, where safety of life or property is involved, a shorter period of notice may be provided. Any protest by a licensee of an FCC order of modification will be handled in accordance with the provisions of 47 U.S.C. §316.

§97.29 Replacement license document.

Each person who has been granted an amateur station license or *reciprocal permit for alien amateur licensee* whose original license document or permit document is lost, mutilated or destroyed must request a replacement. The request must be made to: FCC, 1270 Fairfield Road, Gettysburg, PA 17325-7245. A statement of how the document was lost, mutilated, or destroyed must be attached to the request. A replacement document must bear the same expiration date as the document that it replaces.

Subpart B—Station Operation Standards

§97.101 General standards.

(a) In all respects not specifically covered by FCC Rules each amateur station must be operated in accordance with good engineering and good amateur practice.

(b) Each station licensee and each control operator must cooperate in selecting transmitting channels and in making the most effective use of the amateur service frequencies. No frequency will be assigned for the exclusive use of any station.

(c) At all times and on all frequencies, each control operator must give priority to stations providing emergency communications, except to stations transmitting communications for training drills and tests in RACES.

(d) No amateur operator shall willfully or maliciously interfere with or cause interference to any radio communication or signal.

§97.103 Station licensee responsibilities.

(a) The station licensee is responsible for the proper operation of the station in accordance with the FCC Rules. When the control operator is a different amateur operator than the station licensee, both persons are equally responsible for proper operation of the station.

(b) The station licensee must designate the station control operator. The FCC will presume that the station licensee is also the control operator, unless documentation to the contrary is in the station records.

(c) The station licensee must make the station and the station records available for inspection upon request by an FCC representative. When deemed necessary by an EIC to assure compliance with FCC Rules, the station licensee must maintain a record of station operations containing such items of information as the EIC may require in accord with §0.314(x) of the FCC Rules.

§97.105 Control operator duties.

(a) The control operator must ensure the immediate proper operation of the station, regardless of the type of control.

(b) A station may only be operated in the manner and to the extent permitted by the privileges authorized for the class of operator license held by the control operator.

§97.107 Alien control operator privileges.

(a) The privileges available to a control operator holding an amateur service license issued by the Government of Canada are:

(1) The terms of the *Convention Between the United States and Canada (TIAS no. 2508) Relating to the Operation by Citizens of Either Country of Certain Radio Equipment or Stations in the Other Country;*

(2) The operating terms and conditions of the amateur service license issued by the Government of Canada; and

(3) The applicable provisions of the FCC Rules, but not to exceed the control operator privileges of an FCC-issued Amateur Extra Class operator license.

(b) The privileges available to a control operator holding an FCC-issued reciprocal permit for alien amateur licensee are:

(1) The terms of the agreement between the alien's government and the United States;

(2) The operating terms and conditions of the amateur service license issued by the alien's government;

(3) The applicable provisions of the FCC Rules, but not to exceed the control operator privileges of an FCC-issued Amateur Extra Class operator license; and

(4) None, if the holder of the reciprocal permit has obtained an FCC-issued operator/primary station license.

(c) At any time the FCC may, in its discretion, modify, suspend, or cancel the amateur service privileges within or over any area where radio services are regulated by the FCC of any Canadian amateur service licensee or alien reciprocal permittee.

§97.109 Station control.

(a) Each amateur station must have at least one control point.

(b) When a station is being locally controlled, the control operator must be at the control point. Any station may be locally controlled.

(c) When a station is being remotely controlled, the control operator must be at the control point. Any station may be remotely controlled.

(d) When a station is being automatically controlled, the control operator need not be at the control point. Only stations specifically designated elsewhere in this Part may be automatically controlled. Automatic control must cease upon notification by an EIC that the station is transmitting improperly or causing harmful interference to other stations. Automatic control must not be resumed without prior approval of the EIC.

(e) No station may be automatically controlled while transmitting third party communications, except a station transmitting a RTTY or data emission. All messages that are retransmitted must originate at a station that is being locally or remotely controlled.

§97.111 Authorized transmissions.

(a) An amateur station may transmit the following types of two-way communications:

(1) Transmissions necessary to exchange messages with other stations in the amateur service, except those in any country whose administration has given notice that it objects to such communications. The FCC will issue public notices of current arrangements for international communications;

(2) Transmissions necessary to exchange messages with a station in another FCC-regulated service while providing emergency communications;

(3) Transmissions necessary to exchange messages with a United States government station, necessary to providing communications in RACES; and

(4) Transmissions necessary to exchange messages with a station in a service not regulated by the FCC, but authorized by the FCC to communicate with amateur stations. An amateur station may exchange messages with a participating United States military station during an Armed Forces Day Communications Test.

(b) In addition to one-way transmissions specifically authorized elsewhere in this Part, an amateur station may transmit the following types of one-way communications:

(1) Brief transmissions necessary to make adjustments to the station;

(2) Brief transmissions necessary to establishing two-way communications with other stations;

(3) Telecommand;

(4) Transmissions necessary to providing emergency communications;

(5) Transmissions necessary to assisting persons learning, or improving proficiency in, the international Morse code;

(6) Transmissions necessary to disseminate information bulletins;

(7) Transmissions of telemetry.

§97.113 Prohibited transmissions.

(a) No amateur station shall transmit:

(1) Communications specifically prohibited elsewhere in this Part;

(2) Communications for hire or for material compensation, direct or indirect, paid or promised, except as otherwise provided in these rules;

(3) Communications in which the station licensee or control operator has a pecuniary interest, including communications on behalf of an employer. Amateur operators may, however, notify other amateur operators of the availability for sale or trade of apparatus normally used in an amateur station, provided that such activity is not conducted on a regular basis;

(4) Music using a phone emission except as specifically provided elsewhere in this Section; communications intended to facilitate a criminal act; messages in codes or ciphers intended to obscure the meaning thereof, except as otherwise provided herein; obscene or indecent words or language; or false or deceptive messages, signals or identification;

(5) Communications, on a regular basis, which could reasonably be furnished alternatively through other radio services.

(b) An amateur station shall not engage in any form of broadcasting, nor may an amateur station transmit one-way communications except as specifically provided in these rules; nor shall an amateur station engage in any activity related to program production or news gathering for broadcasting purposes, except that communications directly related to the immediate safety of human life or the protection of property may be provided by amateur stations to broadcasters for dissemination to the public where no other means of communication is reasonably available before or at the time of the event.

(c) A control operator may accept compensation as an incident of a teaching position during periods of time when an amateur station is used by that teacher as a part of classroom instruction at an educational institution.

(d) The control operator of a club station may accept compensation for the periods of time when the station is transmitting telegraphy practice or information bulletins, provided that the station transmits such telegraphy practice and bulletins for at least 40 hours per week; schedules operations on at least six amateur service MF and HF bands using reasonable measures to maximize coverage; where the schedule of normal operating times and frequencies is published at least 30 days in advance of the actual transmissions; and where the control operator does not accept any direct or indirect compensation for any other service as a control operator.

(e) No station shall retransmit programs or signals emanating from any type of radio station other than an amateur station, except propagation and weather forecast information intended for use by the general public and originated from United States Government stations and communications, including incidental music, originating on United States Government frequencies between a space shuttle and its associated Earth stations. Prior approval for shuttle retransmissions must be obtained from the National Aeronautics and Space Administration. Such retransmissions must be for the exclusive use of amateur operators. Propagation, weather forecasts, and shuttle retransmissions may not be conducted on a regular basis, but only occasionally, as an incident of normal amateur radio communications.

(f) No amateur station, except an auxiliary, repeater or space station, may automatically retransmit the radio signals of other amateur stations.

§97.115 Third party communications.

(a) An amateur station may transmit messages for a third party to:
   (1) Any station within the jurisdiction of the United States.
   (2) Any station within the jurisdiction of any foreign government whose administration has made arrangements with the United States to allow amateur stations to be used for transmitting international communications on behalf of third parties. No station shall transmit messages for a third party to any station within the jurisdiction of any foreign government whose administration has not made such an arrangement. This prohibition does not apply to a message for any third party who is eligible to be a control operator of the station.

(b) The third party may participate in stating the message where:
   (1) The control operator is present at the control point and is continuously monitoring and supervising the third party's participation; and
   (2) The third party is not a prior amateur service licensee whose license was revoked; suspended for less than the balance of the license term and the suspension is still in effect; suspended for the balance of the license term and relicensing has not taken place; or surrendered for cancellation following notice of revocation, suspension or monetary forfeiture proceedings. The third party may not be the subject of a cease and desist order which relates to amateur service operation and which is still in effect.

(c) At the end of an exchange of international third party communications, the station must also transmit in the station identification procedure the call sign of the station with which a third party message was exchanged.

§97.117 International communications.

Transmissions to a different country, where permitted, shall be made in plain language and shall be limited to messages of a technical nature relating to tests, and, to remarks of a personal character for which, by reason of their unimportance, recourse to the public telecommunications service is not justified.

§97.119 Station identification.

(a) Each amateur station, except a space station or telecommand station, must transmit its assigned call sign on its transmitting channel at the end of each communication, and at least every ten minutes during a communication, for the purpose of clearly making the source of the transmissions from the station known to those receiving the transmissions. No station may transmit unidentified communications or signals, or transmit as the station call sign, any call sign not authorized to the station.

(b) The call sign must be transmitted with an emission authorized for the transmitting channel in one of the following ways:

(1) By a CW emission. When keyed by an automatic device used only for identification, the speed must not exceed 20 words per minute;

(2) By a phone emission in the English language. Use of a standard phonetic alphabet as an aid for correct station identification is encouraged;

(3) By a RTTY emission using a specified digital code when all or part of the communications are transmitted by a RTTY or data emission;

(4) By an image emission conforming to the applicable transmission standards, either color or monochrome, of §73.682(a) of the FCC Rules when all or part of the communications are transmitted in the same image emission; or

(5) By a CW or phone emission during SS emission transmission on a narrow bandwidth frequency segment. Alternatively, by the changing of one or more parameters of the emission so that a conventional CW or phone emission receiver can be used to determine the station call sign.

(c) An indicator may be included with the call sign. It must be separated from the call sign by the slant mark or by any suitable word that denotes the slant mark. If the indicator is self-assigned it must be included after the call sign and must not conflict with any other indicator specified by the FCC Rules or with any prefix assigned to another country.

(d) When the operator license class held by the control operator exceeds that of the station licensee, an indicator consisting of the call sign assigned to the control operator's station must be included after the call sign.

(e) When the control operator who is exercising the rights and privileges authorized by §97.9(b) of this Part, an indicator must be included after the call sign as follows:

(1) For a control operator who has requested a license modification from Novice to Technician Class: KT;

(2) For a control operator who has requested a license modification from Novice or Technician Class to General Class: AG;

(3) For a control operator who has requested a license modification from

Novice, Technician, or General Class operator to Advanced Class: AA; or

(4) For a control operator who has requested a license modification from Novice, Technician, General, or Advanced Class operator to Amateur Extra Class: AE.

(f) When the station is transmitting under the authority of a reciprocal permit for alien amateur licensee, an indicator consisting of the appropriate letter-numeral designating the station location must be included before the call sign issued to the station by the licensing country. When the station is transmitting under the authority of an amateur service license issued by the Government of Canada, a station location indicator must be included after the call sign. At least once during each intercommunication, the identification announcement must include the geographical location as nearly as possible by city and state, commonwealth or possession.

§97.121 Restricted operation.

(a) If the operation of an amateur station causes general interference to the reception of transmissions from stations operating in the domestic broadcast service when receivers of good engineering design, including adequate selectivity characteristics, are used to receive such transmissions, and this fact is made known to the amateur station licensee, the amateur station shall not be operated during the hours from 8 p.m. to 10:30 p.m., local time, and on Sunday for the additional period from 10:30 a.m. until 1 p.m., local time, upon the frequency or frequencies used when the interference is created.

(b) In general, such steps as may be necessary to minimize interference to stations operating in other services may be required after investigation by the FCC.

Subpart C—Special Operations

§97.201 Auxiliary station.

(a) Any amateur station licensed to a holder of a Technician, General, Advanced or Amateur Extra Class operator license may be an auxiliary station. A holder of a Technician, General, Advanced or Amateur Extra Class operator license may be the control operator of an auxiliary station, subject to the privileges of the class of operator license held.

(b) An auxiliary station may transmit only on the 1.25 m and shorter wavelength bands, except the 219-220 MHz, 222.000-222.150 MHz, 431-433 MHz and 435-438 MHz segments.

(c) Where an auxiliary station causes harmful interference to another auxiliary station, the licensees are equally and fully responsible for resolving the interference unless one station's operation is recommended by a frequency coordinator and the other station's is not. In that case, the licensee of the non-coordinated auxiliary station has primary responsibility to resolve the interference.

(d) An auxiliary station may be automatically controlled.

(e) An auxiliary station may transmit one-way communications.

§97.203 Beacon station.

(a) Any amateur station licensed to a holder of a Technician, General, Advanced or Amateur Extra Class operator license may be a beacon. A holder of a Technician, General, Advanced or Amateur Extra Class operator license may be the control operator of a beacon, subject to the privileges of the class of operator license held.

(b) A beacon must not concurrently transmit on more than 1 channel in the same amateur service frequency band, from the same station location.

(c) The transmitter power of a beacon must not exceed 100 W.

(d) A beacon may be automatically controlled while it is transmitting on the 28.20-28.30 MHz, 50.06-50.08 MHz, 144.275-144.300 MHz, 222.05-222.06 MHz, or 432.300-432.400 MHz segments, or on the 33 cm and shorter wavelength bands.

(e) Before establishing an automatically controlled beacon in the National Radio Quiet Zone or before changing the transmitting frequency, transmitter power, antenna height or directivity, the station licensee must give written notification thereof to the Interference Office, National Radio Astronomy Observatory, P.O. Box 2, Green Bank, WV 24944.

    (1) The notification must include the geographical coordinates of the antenna, antenna ground elevation above mean sea level (AMSL), antenna center of radiation above ground level (AGL), antenna directivity, proposed frequency, type of emission, and transmitter power.

    (2) If an objection to the proposed operation is received by the FCC from the National Radio Astronomy Observatory at Green Bank, Pocahontas County, WV, for itself or on behalf of the Naval Research Laboratory at Sugar Grove, Pendleton County, WV, within 20 days from the date of notification, the FCC will consider all aspects of the problem and take whatever action is deemed appropriate.

(f) A beacon must cease transmissions upon notification by an EIC that the station is operating improperly or causing undue interference to other operations. The beacon may not resume transmitting without prior approval of the EIC.

(g) A beacon may transmit one-way communications.

§97.205 Repeater station.

(a) Any amateur station licensed to a holder of a Technician, General, Advanced or Amateur Extra Class operator license may be a repeater. A holder of a Technician, General, Advanced or Amateur Extra Class operator license may be the control operator of a repeater, subject to the privileges of the class of operator license held.

(b) A repeater may receive and retransmit only on the 10 m and shorter wavelength frequency bands except the 28.0-29.5 MHz, 50.0-51.0 MHz, 144.0-144.5 MHz, 145.5-146.0 MHz, 222.00-222.15 MHz, 431.0-433.0 MHz and 435.0-438.0 MHz segments.

(c) Where the transmissions of a repeater cause harmful interference to another repeater, the two station licensees are equally and fully responsible for resolving the interference unless the operation of one station is recommended by a frequency coordinator and the operation of the other station is not. In that case, the licensee of the noncoordinated repeater has primary responsibility to resolve the interference.

(d) A repeater may be automatically controlled.

(e) Ancillary functions of a repeater that are available to users on the input channel are not considered remotely controlled functions of the station. Limiting the use of a repeater to only certain user stations is permissible.

(f) Before establishing a repeater in the National Radio Quiet Zone or before changing the transmitting frequency, transmitter power, antenna height or directivity, or the location of an existing repeater, the station licensee must give written notification thereof to the Interference Office, National Radio Astronomy Observatory, P.O. Box 2, Green Bank, WV 24944.

  (1) The notification must include the geographical coordinates of the station antenna, antenna ground elevation above mean sea level (AMSL), antenna center of radiation above ground level (AGL), antenna directivity, proposed frequency, type of emission, and transmitter power.

  (2) If an objection to the proposed operation is received by the FCC from the National Radio Astronomy Observatory at Green Bank, Pocahontas County, WV, for itself or on behalf of the Naval Research Laboratory at Sugar Grove, Pendleton County, WV, within 20 days from the date of notification, the FCC will consider all aspects of the problem and take whatever action is deemed appropriate.

(g) The control operator of a repeater that retransmits inadvertently communiations that violate the rules in this Part is not accountable for the violative communications.

§97.207 Space station.

(a) Any amateur station may be a space station. A holder of any class operator license may be the control operator of a space station, subject to the privileges of the class of operator license held by the control operator.

(b) A space station must be capable of effecting a cessation of transmissions by telecommand whenever such cessation is ordered by the FCC.

(c) The following frequency bands and segments are authorized to space stations:

  (1) The 17 m, 15 m, 12 m and 10 m bands, 6 mm, 4 mm, 2 mm and 1 mm bands; and

  (2) The 7.0-7.1 MHz, 14.00-14.25 MHz, 144-146 MHz, 435-438 MHz, 1260-1270 MHz and 2400-2450 MHz, 3.40-3.41 GHz, 5.83-5.85 GHz, 10.45-10.50 GHz and 24.00-24.05 GHz segments.

(d) A space station may automatically retransmit the radio signals of Earth stations and other space stations.

(e) A space station may transmit one-way communications.

(f) Space telemetry transmissions may consist of specially coded messages intended to facilitate communications or related to the function of the spacecraft.

(g) The licensee of each space station must give two written, pre-space station notifications to the Wireless Telecommunications Bureau, FCC, Washington, DC 20554. Each notification must be in accord with the provisions of Articles 11 and 13 of the Radio Regulations.

  (1) The first notification is required no less than 27 months prior to initiating space station transmissions and must specify the information required by Appendix 4, and Resolution No. 642 of the Radio Regulations.

  (2) The second notification is required no less than 5 months prior to initiating space station transmissions and must specify the information re-

quired by Appendix 3 and Resolution No. 642 of the Radio Regulations.

(h) The licensee of each space station must give a written, in-space station notification to the Wireless Telecommunications Bureau, FCC, Washington, DC 20554, no later than 7 days following initiation of space station transmissions. The notification must update the information contained in the pre-space notification.

(i) The licensee of each space station must give a written, post-space notification to the Wireless Telecommunications Bureau, FCC, Washington, DC 20554, no later than 3 months after termination of the space station transmissions. When the termination is ordered by the FCC, notification is required no later than 24 hours after termination.

§97.209 Earth station.

(a) Any amateur station may be an Earth station. A holder of any class operator license may be the control operator of an Earth station, subject to the privileges of the class of operator license held by the control operator.

(b) The following frequency bands and segments are authorized to Earth stations:
(1) The 17 m, 15 m, 12 m and 10 m bands, 6 mm, 4 mm, 2 mm and 1 mm bands; and
(2) The 7.0-7.1 MHz, 14.00-14.25 MHz, 144-146 MHz, 435-438 MHz, 1260-1270 MHz and 2400-2450 MHz, 3.40-3.41 GHz, 5.65-5.67 GHz, 10.45-10.50 GHz and 24.00-24.05 GHz segments.

§97.211 Space Telecommand station.

(a) Any amateur station designated by the licensee of a space station is eligible to transmit as a telecommand station for that space station, subject to the privileges of the class of operator license held by the control operator.

(b) A telecommand station may transmit special codes intended to obscure the meaning of telecommand messages to the station in space operation.

(c) The following frequency bands and segments are authorized to telecommand stations:
(1) The 17 m, 15 m, 12 m and 10 m bands, 6 mm, 4 mm, 2 mm and 1 mm bands; and
(2) The 7.0-7.1 MHz, 14.00-14.25 MHz, 144-146 MHz, 435-438 MHz, 1260-1270 MHz and 2400-2450 MHz, 3.40-3.41 GHz, 5.65-5.67 GHz, 10.45-10.50 GHz and 24.00-24.05 GHz segments.

(d) A telecommand station may transmit one-way communications.

§97.213 Telecommand of an amateur station.

An amateur station on or within 50 km of the Earth's surface may be under telecommand where:

(a) There is a radio or wireline control link between the control point and the station sufficient for the control operator to perform his/her duties. If radio, the control link must use an auxiliary station. A control link using a fiber optic cable or another telecommunication service is considered wireline.

(b) Provisions are incorporated to limit transmission by the station to a period of no more than 3 minutes in the event of malfunction in the control link.

(c) The station is protected against making, willfully or negligently, unauthorized transmissions.

(d) A photocopy of the station license and a label with the name, address, and telephone number of the station licensee and at least one designated control operator is posted in a conspicuous place at the station location.

§97.215 Telecommand of model craft.

An amateur station transmitting signals to control a model craft may be operated as follows:

(a) The station identification procedure is not required for transmissions directed only to the model craft, provided that a label indicating the station call sign and the station licensee's name and address is affixed to the station transmitter.

(b) The control signals are not considered codes or ciphers intended to obscure the meaning of the communication.

(c) The transmitter power must not exceed 1 W.

§97.217 Telemetry.

Telemetry transmitted by an amateur station on or within 50 km of the Earth's surface is not considered to be codes or ciphers intended to obscure the meaning of communications.

§97.219 Message forwarding system.

(a) Any amateur station may participate in a message forwarding system, subject to the privileges of the class of operator license held.

(b) For stations participating in a message forwarding system, the control operator of the station originating a message is primarily accountable for any violation of the rules in this Part contained in the message.

(c) Except as noted in paragraph (d) of this section, for stations participating in a message forwarding system, the control operators of forwarding stations that retransmit inadvertently communications that violate the rules in this Part are not accountable for the violative communications. They are, however, responsible for discontinuing such communications once they become aware of their presence.

(d) For stations participating in a message forwarding system, the control operator of the first forwarding station must:

(1) Authenticate the identity of the station from which it accepts communication on behalf of the system; or

(2) Accept accountability for any violation of the rules in this Part contained in messages it retransmits to the system.

§97.221 Automatically controlled digital station.

(a) This rule section does not apply to an auxiliary station, a beacon station, a repeater station, an earth station, a space station, or a space telecommand station.

(b) A station may be automatically controlled while transmitting a RTTY or data emission on the 6 m or shorter wavelength bands, and on the 28.120-28.189 MHz, 24.925-24.930 MHz, 21.090-21.100 MHz, 18.105-18.110 MHz, 14.0950-14.0995 MHz, 14.1005-14.112 MHz, 10.140-10.150 MHz, 7.100-7.105 MHz, or 3.620-3.635 MHz segments.

(c) A station may be automatically controlled while transmitting a RTTY or data emission on any other frequency authorized for such emission types provided that:

(1) The station is responding to interrogation by a station under local or remote control; and

(2) No transmission from the automatically controlled station occupies a bandwidth of more than 500 Hz.

Subpart D—Technical Standards

§97.301 Authorized frequency bands.

The following transmitting frequency bands are available to an amateur station located within 50 km of the Earth's surface, within the specified ITU Region, and outside any area where the amateur service is regulated by any authority other than the FCC.

(a) For a station having a control operator who has been granted an operator license of Technician, Technician Plus, General, Advanced, or Amateur Extra Class:

Wavelength band	ITU Region 1	ITU Region 2	ITU Region 3	Sharing requirements See §97.303, Paragraph:
VHF	*MHz*	*MHz*	*MHz*	
6 m	—	50-54	50-54	(a)
2 m	144-146	144-148	144-148	(a)
1.25 m	—	219-220	—	(a), (e)
-do-	—	222-225	—	(a)
UHF	*MHz*	*MHz*	*MHz*	
70 cm	430-440	420-450	420-450	(a), (b), (f)
33 cm	—	902-928	—	(a), (b), (g)
23 cm	1240-1300	1240-1300	1240-1300	(h), (i)
13 cm	2300-2310	2300-2310	2300-2310	(a), (b), (j)
-do-	2390-2450	2390-2450	2390-2450	(a), (b), (j)
SHF	*GHz*	*GHz*	*GHz*	
9 cm	—	3.3-3.5	3.3-3.5	(a), (b), (k), (l)
5 cm	5.650-5.850	5.650-5.925	5.650-5.850	(a), (b), (m)
3 cm	10.00-10.50	10.00-10.50	10.00-10.50	(b), (c), (i), (n)
1.2 cm	24.00-24.25	24.00-24.25	24.00-24.25	(a), (b), (h), (o)
EHF	*GHz*	*GHz*	*GHz*	
6 mm	47.0-47.2	47.0-47.2	47.0-47.2	
4 mm	75.5-81.0	75.5-81.0	75.5-81.0	(b), (c), (h)
2.5 mm	119.98-120.02	119.98-120.02	119.98-120.02	(k), (p)
2 mm	142-149	142-149	142-149	(b), (c), (h), (k)
1 mm	241-250	241-250	241-250	(b), (c), (h), (q)
—	above 300	above 300	above 300	(k)

(b) For a station having a control operator who has been granted an operator license of Amateur Extra Class:

Wavelength band	ITU Region 1	ITU Region 2	ITU Region 3	Sharing requirements See §97.303, Paragraph:
MF	*kHz*	*kHz*	*kHz*	
160 m	1810-1850	1800-2000	1800-2000	(a), (b), (c)
HF	*MHz*	*MHz*	*MHz*	
80 m	3.50-3.75	3.50-3.75	3.50-3.75	(a)
75 m	3.75-3.80	3.75-4.00	3.75-3.90	(a)
40 m	7.0-7.1	7.0-7.3	7.0-7.1	(a)
30 m	10.10-10.15	10.10-10.15	10.10-10.15	(d)
20 m	14.00-14.35	14.00-14.35	14.00-14.35	
17 m	18.068-18.168	18.068-18.168	18.068-18.168	
15 m	21.00-21.45	21.00-21.45	21.00-21.45	
12 m	24.89-24.99	24.89-24.99	24.89-24.99	
10 m	28.0-29.7	28.0-29.7	28.0-29.7	

(c) For a station having a control operator who has been granted an operator license of Advanced Class:

Wavelength band	ITU Region 1	ITU Region 2	ITU Region 3	Sharing requirements See §97.303, Paragraph:
MF	*kHz*	*kHz*	*kHz*	
160 m	1810-1850	1800-2000	1800-2000	(a), (b), (c)
HF	*MHz*	*MHz*	*MHz*	
80 m	3.525-3.750	3.525-3.750	3.525-3.750	(a)
75 m	3.775-3.800	3.775-4.000	3.775-3.900	(a)
40 m	7.025-7.100	7.025-7.300	7.025-7.100	(a)
30 m	10.10-10.15	10.10-10.15	10.10-10.15	(d)
20 m	14.025-14.150	14.025-14.150	14.025-14.150	
-do-	14.175-14.350	14.175-14.350	14.175-14.350	
17 m	18.068-18.168	18.068-18.168	18.068-18.168	
15 m	21.025-21.200	21.025-21.200	21.025-21.200	
-do-	21.225-21.450	21.225-21.450	21.225-21.450	
12 m	24.89-24.99	24.89-24.99	24.89-24.99	
10 m	28.0-29.7	28.0-29.7	28.0-29.7	

(d) For a station having a control operator who has been granted an operator license of General Class:

Wavelength band	ITU Region 1	ITU Region 2	ITU Region 3	Sharing requirements See §97.303, Paragraph:
MF	*kHz*	*kHz*	*kHz*	
160 m	1810-1850	1800-2000	1800-2000	(a), (b), (c)
HF	*MHz*	*MHz*	*MHz*	
80 m	3.525-3.750	3.525-3.750	3.525-3.750	(a)
75 m	—	3.85-4.00	3.85-3.90	(a)
40 m	7.025-7.100	7.025-7.150	7.025-7.100	(a)
-do-	—	7.225-7.300	—	(a)
30 m	10.10-10.15	10.10-10.15	10.10-10.15	(d)
20 m	14.025-14.150	14.025-14.150	14.025-14.150	
-do-	14.225-14.350	14.225-14.350	14.225-14.350	
17 m	18.068-18.168	18.068-18.168	18.068-18.168	
15 m	21.025-21.200	21.025-21.200	21.025-21.200	
-do-	21.30-21.45	21.30-21.45	21.30-21.45	
12 m	24.89-24.99	24.89-24.99	24.89-24.99	
10 m	28.0-29.7	28.0-29.7	28.0-29.7	

(e) For a station having a control operator who has been granted an operator license of Novice or Technician Plus Class:

Wavelength band	ITU Region 1	ITU Region 2	ITU Region 3	Sharing requirements See §97.303, Paragraph:
HF	*MHz*	*MHz*	*MHz*	
80 m	3.675-3.725	3.675-3.725	3.675-3.725	(a)
40 m	7.050-7.075	7.10-7.15	7.050-7.075	(a)
15 m	21.10-21.20	21.10-21.20	21.10-21.20	
10 m	28.1-28.5	28.1-28.5	28.1-28.5	

(f) For a station having a control operator who has been granted an operator license of Novice Class:

Wavelength band	ITU Region 1	ITU Region 2	ITU Region 3	Sharing requirements See §97.303, Paragraph:
VHF	MHz	MHz	MHz	
1.25 m	—	222-225	—	(a)
UHF	*MHz*	*MHz*	*MHz*	
23 cm	1270-1295	1270-1295	1270-1295	(h), (i)

§97.303 Frequency sharing requirements.

The following is a summary of the frequency sharing requirements that apply to amateur station transmissions on the frequency bands specified in §97.301 of this Part. (For each ITU Region, each frequency band allocated to the amateur service is designated as either a secondary service or a primary service. A station in a secondary service must not cause harmful interference to, and must accept interference from, stations in a primary service. See §§2.105 and 2.106 of the FCC Rules, *United States Table of Frequency Allocations* for complete requirements.)

(a) Where, in adjacent ITU Regions or Subregions, a band of frequencies is allocated to different services of the same category, the basic principle is the equality of right to operate. The stations of each service in one region must operate so as not to cause harmful interference to services in the other Regions or Subregions. (See ITU *Radio Regulations*, No. 346 (Geneva, 1979).)

(b) No amateur station transmitting in the 1900-2000 kHz segment, the 70 cm band, the 33 cm band, the 13 cm band, the 9 cm band, the 5 cm band, the 3 cm band, the 24.05-24.25 GHz segment, the 76-81 GHz segment, the 144-149 GHz segment and the 241-248 GHz segment shall cause harmful interference to, nor is protected from interference due to the operation of, the Government radiolocation service.

(c) No amateur station transmitting in the 1900-2000 kHz segment, the 3 cm band, the 76-81 GHz segment, the 144-149 GHz segment and the 241-248 GHz segment shall cause harmful interference to, nor is protected from interference due to the operation of, stations in the non-Government radiolocation service.

(d) No amateur station transmitting in the 30 meter band shall cause harmful interference to stations authorized by other nations in the fixed service. The licensee of the amateur station must make all necessary adjustments, including termination of transmissions, if harmful interference is caused.

(e) In the 1.25 m band:

(1) Use of the 219-220 MHz segment is limited to amateur stations participating, as forwarding stations, in point-to-point fixed digital message forwarding systems, including intercity packet backbone networks. It is not available for other purposes.

(2) No amateur station transmitting in the 219-220 MHz segment shall cause harmful interference to, nor is protected from interference due to operation of Automated Maritime Telecommunications Systems (AMTS), television broadcasting on channels 11 and 13, Interactive Video and Data Service systems, Land Mobile Services systems, or any other service having a primary allocation in or adjacent to the band.

(3) No amateur station may transmit in the 219-220 MHz segment unless the licensee has given written notification of the station's specific geographic location for such transmissions in order to be incorporated into a data base that has been made available to the public. The notification must be given at least 30 days prior to making such transmissions. The notification must be given to:

> *The American Radio Relay League*
> *225 Main Street*
> *Newington, CT 06111-1494*

(4) No amateur station may transmit in the 219-220 MHz segment from a location that is within 640 km of an AMTS Coast Station that uses frequencies in the 217-218/219-220 MHz AMTS bands unless the amateur station licensee has given written notification of the station's specific geographic location for such transmissions to the AMTS licensee.

The notification must be given at least 30 days prior to making such transmissions. The location of AMTS Coast Stations using the 217-218/219-220 MHz channels may be obtained from either:

The American Radio Relay League
225 Main Street
Newington, CT 06111-1494

or

Interactive Systems, Inc.
Suite 1103
1601 North Kent Street
Arlington, VA 22209
Fax: (703) 812-8275
Phone: (703) 812-8270

(5) No amateur station may transmit in the 219-220 MHz segment from a location that is within 80 km of an AMTS Coast Station that uses frequencies in the 217-218/219-220 MHz AMTS bands unless that amateur station licensee holds written approval from that AMTS licensee. The location of AMTS Coast Stations using the 217-218/219-220 MHz channels may be obtained as noted in paragragh (e)(4) of this section.

(f) In the 70 cm band:
    (1) No amateur station shall transmit from north of Line A in the 420-430 MHz segment.

    (2) The 420-430 MHz segment is allocated to the amateur service in the United States on a secondary basis, and is allocated in the fixed and mobile (except aeronautical mobile) services in the International Table of allocations on a primary basis. No amateur station transmitting in this band shall cause harmful interference to, nor is protected from interference due to the operation of, stations authorized by other nations in the fixed and mobile (except aeronautical mobile) services.

    (3) The 430-440 MHz segment is allocated to the amateur service on a secondary basis in ITU Regions 2 and 3. No amateur station transmitting in this band in ITU Regions 2 and 3 shall cause harmful interference to, nor is protected from interference due to the operation of, stations authorized by other nations in the radiolocation service. In ITU Region 1, the 430-440 MHz segment is allocated to the amateur service on a co-primary basis with the radiolocation service. As between these two services in this band in ITU Region 1, the basic principle that applies is the equality of right to operate. Amateur stations authorized by the United States and radiolocation stations authorized by other nations in ITU Region 1 shall operate so as not to cause harmful interference to each other.

    (4) No amateur station transmitting in the 449.75-450.25 MHz segment shall cause interference to, nor is protected from interference due to the operation of stations in, the space operation service and the space research service or Government or non-Government stations for space telecommand.

(g) In the 33 cm band:
    (1) No amateur station shall transmit from within the States of Colorado

and Wyoming, bounded on the south by latitude 39° N, on the north by latitude 42° N, on the east by longitude 105° W, and on the west by longitude 108° W.[1] This band is allocated on a secondary basis to the amateur service subject to not causing harmful interference to, and not receiving protection from any interference due to the operation of, industrial, scientific and medical devices, automatic vehicle monitoring systems or Government stations authorized in this band.

(2) No amateur station shall transmit from those portions of the States of Texas and New Mexico bounded on the south by latitude 31° 41' N, on the north by latitude 34° 30' N, on the east by longitude 104° 11' W, and on the west by longitude 107° 30' W.

(h) No amateur station transmitting in the 23 cm band, the 3 cm band, the 24.05-24.25 GHz segment, the 76-81 GHz segment, the 144-149 GHz segment and the 241-248 GHz segment shall cause harmful interference to, nor is protected from interference due to the operation of, stations authorized by other nations in the radiolocation service.

(i) In the 1240-1260 MHz segment, no amateur station shall cause harmful interference to, nor is protected from interference due to the operation of, stations in the radionavigation-satellite service, the aeronautical radio-navigation service, or the radiolocation service.

(j) In the 13 cm band:

(1) The amateur service is allocated on a secondary basis in all ITU Regions. In ITU Region 1, no amateur station shall cause harmful interference to, and is not protected from interference due to the operation of, stations authorized by other nations in the fixed service. In ITU Regions 2 and 3, no station shall cause harmful interference to, and is not protected from interference due to the operation of, stations authorized by other nations in the fixed, mobile and radiolocation services.

(2) In the United States, the 2300-2310 MHz segment is allocated to the amateur service on a co-secondary basis with the Government fixed and mobile services. In this segment, the fixed and mobile services must not cause harmful interference to the amateur service. No amateur station transmitting in the 2400-2450 MHz segment is protected from interference due to the operation of industrial, scientific and medical devices on 2450 MHz.

(k) No amateur station transmitting in the 3.332-3.339 GHz and 3.3458-3525 GHz segments, the 2.5 mm band, the 144.68-144.98 GHz, 145.45-145.75 GHz and 146.82-147.12 GHz segments and the 343-348 GHz segment shall cause harmful interference to stations in the radio astronomy service. No amateur station transmitting in the 300-302 GHz, 324-326 GHz, 345-347 GHz, 363-365 GHz and 379-381 GHz segments shall cause harmful interference to stations in the space research service (passive) or Earth exploration-satellite service (passive).

(l) In the 9 cm band:

(1) In ITU Regions 2 and 3, the band is allocated to the amateur service on a secondary basis.

[1] In a waiver effective July 2, 1990, the FCC permitted amateurs in the restricted areas to transmit in the following segments: 902.0-902.4, 902.6-904.3, 904.7-925.3, 925.7-927.3, and 927.7-928.0 MHz.

(2) In the United States, the band is allocated to the amateur service on a co-secondary basis with the non-Government radiolocation service.

(3) In the 3.3-3.4 GHz segment, no amateur station shall cause harmful interference to, nor is protected from interference due to the operation of, stations authorized by other nations in the fixed and fixed-satellite service.

(4) In the 3.4-3.5 GHz segment, no amateur station shall cause harmful interference to, nor is protected from interference due to the operation of, stations authorized by other nations in the fixed and fixed-satellite service.

(m) In the 5 cm band:

(1) In the 5.650-5.725 GHz segment, the amateur service is allocated in all ITU Regions on a co-secondary basis with the space research (deep space) service.

(2) In the 5.725-5.850 GHz segment, the amateur service is allocated in all ITU Regions on a secondary basis. No amateur station shall cause harmful interference to, nor is protected from interference due to the operation of, stations authorized by other nations in the fixed-satellite service in ITU Region 1.

(3) No amateur station transmitting in the 5.725-5.875 GHz segment is protected from interference due to the operation of industrial, scientific and medical devices operating on 5.8 GHz.

(4) In the 5.650-5.850 GHz segment, no amateur station shall cause harmful interference to, nor is protected from interference due to the operation of, stations authorized by other nations in the radiolocation service.

(5) In the 5.850-5.925 GHz segment, the amateur service is allocated in ITU Region 2 on a co-secondary basis with the radiolocation service. In the United States, the segment is allocated to the amateur service on a secondary basis to the non-Government fixed-satellite service. No amateur station shall cause harmful interference to, nor is protected from interference due to the operation of, stations authorized by other nations in the fixed, fixed-satellite and mobile services. No amateur station shall cause harmful interference to, nor is protected from interference due to the operation of, stations in the non-Government fixed-satellite service.

(n) In the 3 cm band:

(1) In the United States, the 3 cm band is allocated to the amateur service on a co-secondary basis with the non-government radiolocation service.

(2) In the 10.00-10.45 GHz segment in ITU Regions 1 and 3, no amateur station shall cause interference to, nor is protected from interference due to the operation of, stations authorized by other nations in the fixed and mobile services.

(o) No amateur station transmitting in the 1.2 cm band is protected from interference due to the operation of industrial, scientific and medical devices on 24.125 GHz. In the United States, the 24.05-24.25 GHz segment is allocated to the amateur service on a co-secondary basis with the non-government radiolocation and Government and non-government Earth exploration-satellite (active) services.

(p) The 2.5 mm band is allocated to the amateur service on a secondary basis. No amateur station transmitting in this band shall cause harmful interference to, nor is protected from interference due to the operation of, stations in the fixed, inter-satellite and mobile services.

(q) No amateur station transmitting in the 244-246 GHz segment of the 1 mm band is protected from interference due to the operation of industrial, scientific and medical devices on 245 GHz.

§97.305 Authorized emission types.

(a) An amateur station may transmit a CW emission on any frequency authorized to the control operator.

(b) A station may transmit a test emission on any frequency authorized to the control operator for brief periods for experimental purposes, except that no pulse modulation emission may be transmitted on any frequency where pulse is not specifically authorized.

(c) A station may transmit the following emission types on the frequencies indicated, as authorized to the control operator, subject to the standards specified in §97.307(f) of this part.

Wavelength band	Frequencies	Emission Types Authorized	Standards See §97.307(f), paragraph:
MF:			
160 m	Entire band	RTTY, data	(3)
-do-	-do-	Phone, image	(1), (2)
HF:			
80 m	Entire band	RTTY, data	(3), (9)
75 m	Entire band	Phone, image	(1), (2)
40 m	7.000-7.100 MHz	RTTY, data	(3), (9)
-do-	7.075-7.100 MHz	Phone, image	(1), (2), (9), (11)
-do-	7.100-7.150 MHz	RTTY, data	(3), (9)
-do-	7.150-7.300 MHz	Phone, image	(1), (2)
30 m	Entire band	RTTY, data	(3)
20 m	14.00-14.15 MHz	RTTY, data	(3)
-do-	14.15-14.35 MHz	Phone, image	(1), (2)
17 m	18.068-18.110 MHz	RTTY, data	(3)
-do-	18.110-18.168 MHz	Phone, image	(1), (2)
15 m	21.0-21.2 MHz	RTTY, data	(3), (9)
-do-	21.20-21.45 MHz	Phone, image	(1), (2)
12 m	24.89-24.93 MHz	RTTY, data	(3)
-do-	24.93-24.99 MHz	Phone, image	(1), (2)
10 m	28.0-28.3 MHz	RTTY, data	(4)
-do-	28.3-28.5 MHz	Phone, image	(1), (2), (10)
-do-	28.5-29.0 MHz	Phone, image	(1), (2)
-do-	29.0-29.7MHz	Phone, image	(2)
VHF:			
6 m	50.1-51.0 MHz	RTTY, data	(5)
-do-	-do-	MCW, phone, image	(2)
-do-	51.0-54.0 MHz	RTTY, data, test	(5), (8)
-do-	-do-	MCW, phone, image	(2)

Wavelength band	Frequencies	Emission Types Authorized	Standards See §97.307(f), paragraph:
2 m	144.1-148.0 MHz	RTTY, data, test	(5), (8)
-do-	-do-	MCW, phone, image	(2)
1.25 m	219-220 MHz	Data	(13)
-do-	222-225 MHz	MCW, phone, image RTTY, data, test	(2), (6), (8)

UHF:

Wavelength band	Frequencies	Emission Types Authorized	Standards
70 cm	Entire band	MCW, phone, image, RTTY, data, SS, test	(6), (8)
33 cm	Entire band	MCW, phone, image, RTTY, data, SS, test, pulse	(7), (8), (12)
23 cm	Entire band	MCW, phone, image, RTTY, data, SS, test	(7), (8), (12)
13 cm	Entire band	MCW, phone, image, RTTY, data, SS, test, pulse	(7), (8), (12)

SHF:

Wavelength band	Frequencies	Emission Types Authorized	Standards
9 cm	Entire band	MCW, phone, image, RTTY, data, SS, test, pulse	(7), (8), (12)
5 cm	Entire band	MCW, phone, image, RTTY, data, SS, test, pulse	(7), (8), (12)
3 cm	Entire band	MCW, phone, image, RTTY, data, SS, test	(7), (8), (12)
1.2 cm	Entire band	MCW, phone, image, RTTY, data, SS, test, pulse	(7), (8), (12)

EHF:

Wavelength band	Frequencies	Emission Types Authorized	Standards
6 mm	Entire band	MCW, phone, image, RTTY, data, SS, test, pulse	(7), (8), (12)
4 mm	Entire band	MCW, phone, image, RTTY, data, SS, test, pulse	(7), (8), (12)
2.5 mm	Entire band	MCW, phone, image, RTTY, data, SS, test, pulse	(7), (8), (12)
2 mm	Entire band	MCW, phone, image, RTTY, data, SS, test, pulse	(7), (8), (12)
1 mm	Entire band	MCW, phone, image, RTTY, data, SS, test, pulse	(7), (8), (12)

Wavelength band	Frequencies	Emission Types Authorized	Standards See §97.307(f), paragraph:
—	Above 300 GHz	MCW, phone, image, RTTY, data, SS, test, pulse	(7), (8), (12)

§97.307 Emission standards.

(a) No amateur station transmission shall occupy more bandwidth than necessary for the information rate and emission type being transmitted, in accordance with good amateur practice.

(b) Emissions resulting from modulation must be confined to the band or segment available to the control operator. Emissions outside the necessary bandwidth must not cause splatter or keyclick interference to operations on adjacent frequencies.

(c) All spurious emissions from a station transmitter must be reduced to the greatest extent practicable. If any spurious emission, including chassis or power line radiation, causes harmful interference to the reception of another radio station, the licensee of the interfering amateur station is required to take steps to eliminate the interference, in accordance with good engineering practice.

(d) The mean power of any spurious emission from a station transmitter or external RF power amplifier transmitting on a frequency below 30 MHz must not exceed 50 mW and must be at least 40 dB below the mean power of the fundamental emission. For a transmitter of mean power less than 5 W, the attenuation must be at least 30 dB. A transmitter built before April 15, 1977, or first marketed before January 1, 1978, is exempt from this requirement.

(e) The mean power of any spurious emission from a station transmitter or external RF power amplifier transmitting on a frequency between 30-225 MHz must be at least 60 dB below the mean power of the fundamental. For a transmitter having a mean power of 25 W or less, the mean power of any spurious emission supplied to the antenna transmission line must not exceed 25 μW and must be at least 40 dB below the mean power of the fundamental emission, but need not be reduced below the power of 10 μW. A transmitter built before April 15, 1977, or first marketed before January 1, 1978, is exempt from this requirement.

(f) The following standards and limitations apply to transmissions on the frequencies specified in §97.305(c) of this Part.

    (1) No angle-modulated emission may have a modulation index greater than 1 at the highest modulation frequency.

    (2) No non-phone emission shall exceed the bandwidth of a communications quality phone emission of the same modulation type. The total bandwidth of an independent sideband emission (having B as the first symbol), or a multiplexed image and phone emission, shall not exceed that of a communications quality A3E emission.

    (3) Only a RTTY or data emission using a specified digital code listed in §97.309(a) of this Part may be transmitted. The symbol rate must not exceed 300 bauds, or for frequency-shift keying, the frequency shift between mark and space must not exceed 1 kHz.

(4) Only a RTTY or data emission using a specified digital code listed in §97.309(a) of this Part may be transmitted. The symbol rate must not exceed 1200 bauds. For frequency-shift keying, the frequency shift between mark and space must not exceed 1 kHz.

(5) A RTTY, data or multiplexed emission using a specified digital code listed in §97.309(a) of this Part may be transmitted. The symbol rate must not exceed 19.6 kilobauds. A RTTY, data or multiplexed emission using an unspecified digital code under the limitations listed in §97.309(b) of this Part also may be transmitted. The authorized bandwidth is 20 kHz.

(6) A RTTY, data or multiplexed emission using a specified digital code listed in §97.309(a) of this Part may be transmitted. The symbol rate must not exceed 56 kilobauds. A RTTY, data or multiplexed emission using an unspecified digital code under the limitations listed in §97.309(b) of this Part also may be transmitted. The authorized bandwidth is 100 kHz.

(7) A RTTY, data or multiplexed emission using a specified digital code listed in §97.309(a) of this Part or an unspecified digital code under the limitations listed in §97.309(b) of this Part may be transmitted.

(8) A RTTY or data emission having designators with A, B, C, D, E, F, G, H, J or R as the first symbol; 1, 2, 7 or 9 as the second symbol; and D or W as the third symbol is also authorized.

(9) A station having a control operator holding a Novice or Technician Class operator license may only transmit a CW emission using the international Morse code.

(10) A station having a control operator holding a Novice or Technician Class operator license may only transmit a CW emission using the international Morse code or phone emissions J3E and R3E.

(11) Phone and image emissions may be transmitted only by stations located in ITU Regions 1 and 3, and by stations located within ITU Region 2 that are west of 130° West longitude or south of 20° North latitude.

(12) Emission F8E may be transmitted.

(13) A data emission using an unspecified digital code under the limitations listed in § 97.309(b) of this Part also may be transmitted. The authorized bandwidth is 100 kHz.

§97.309 RTTY and data emission codes.

(a) Where authorized by §97.305(c) and 97.307(f) of this Part, an amateur station may transmit a RTTY or data emission using the following specified digital codes:

(1) The 5-unit, start-stop, International Telegraph Alphabet No. 2, code defined in International Telegraph and Telephone Consultative Committee Recommendation F.1, Division C (commonly known as Baudot).

(2) The 7-unit code, specified in International Radio Consultative Committee Recommendation CCIR 476-2 (1978), 476-3 (1982), 476-4 (1986) or 625 (1986) (commonly known as AMTOR).

(3) The 7-unit code defined in American National Standards Institute X3.4-1977 or International Alphabet No. 5 defined in International Telegraph and Telephone Consultative Committee Recommendation T.50 or in International Organization for Standardization, International Standard ISO 646 (1983), and extensions as provided for in CCITT Recommenda-

tion T.61 (Malaga-Torremolinos, 1984) (commonly known as ASCII).

(4) An amateur station transmitting a RTTY or data emission using a digital code specified in this paragraph may use any technique whose technical characteristics have been documented publicly, such as CLOVER, G-TOR, or PacTOR, for the purpose of facilitating communications.

(b) Where authorized by §§97.305(c) and 97.307(f) of this Part, a station may transmit a RTTY or data emission using an unspecified digital code, except to a station in a country with which the United States does not have an agreement permitting the code to be used. RTTY and data emissions using unspecified digital codes must not be transmitted for the purpose of obscuring the meaning of any communication. When deemed necessary by an EIC to assure compliance with the FCC Rules, a station must:

(1) Cease the transmission using the unspecified digital code;

(2) Restrict transmissions of any digital code to the extent instructed;

(3) Maintain a record, convertible to the original information, of all digital communications transmitted.

§97.311 SS emission types.

(a) SS emission transmissions by an amateur station are authorized only for communications between points within areas where the amateur service is regulated by the FCC. SS emission transmissions must not be used for the purpose of obscuring the meaning of any communication.

(b) Stations transmitting SS emission must not cause harmful interference to stations employing other authorized emissions, and must accept all interference caused by stations employing other authorized emissions. For the purposes of this paragraph, unintended triggering of carrier operated repeaters is not considered to be harmful interference.

(c) Only the following types of SS emission transmissions are authorized (hybrid SS emission transmissions involving both spreading techniques are prohibited):

(1) Frequency hopping where the carrier of the transmitted signal is modulated with unciphered information and changes frequency at fixed intervals under the direction of a high speed code sequence.

(2) Direct sequence where the information is modulo-2 added to a high speed code sequence. The combined information and code are then used to modulate the RF carrier. The high speed code sequence dominates the modulation function, and is the direct cause of the wide spreading of the transmitted signal.

(d) The only spreading sequences that are authorized are from the output of one binary linear feedback shift register (which may be implemented in hardware or software).

(1) Only the following sets of connections may be used:

Number of stages in shift register	Taps used in feedback
7	7, 1.
13	13, 4, 3, and 1.
19	19, 5, 2, and 1.

(2) The shift register must not be reset other than by its feedback during an individual transmission. The shift register output sequence must be

used without alteration.

(3) The output of the last stage of the binary linear feedback shift register must be used as follows:

(i) For frequency hopping transmissions using x frequencies, n consecutive bits from the shift register must be used to select the next frequency from a list of frequencies sorted in ascending order. Each consecutive frequency must be selected by a consecutive block of n bits. (Where n is the smallest integer greater than $\log_2 X$.)

(ii) For direct sequence transmissions using m-ary modulation, consecutive blocks of $\log_2$ m bits from the shift register must be used to select the transmitted signal during each interval.

(e) The station records must document all SS emission transmissions and must be retained for a period of 1 year following the last entry. The station records must include sufficient information to enable the FCC, using the information contained therein, to demodulate all transmissions. The station records must contain at least the following:

(1) A technical description of the transmitted signal;

(2) Pertinent parameters describing the transmitted signal including the frequency or frequencies of operation and, where applicable, the chip rate, the code rate, the spreading function, the transmission protocol(s) including the method of achieving synchronization, and the modulation type;

(3) A general description of the type of information being conveyed (voice, text, memory dump, facsimile, television, etc.);

(4) The method and, if applicable, the frequency or frequencies used for station identification; and

(5) The date of beginning and the date of ending use of each type of transmitted signal.

(f) When deemed necessary by an EIC to assure compliance with this Part, a station licensee must:

(1) Cease SS emission transmissions;

(2) Restrict SS emission transmissions to the extent instructed; and

(3) Maintain a record, convertible to the original information (voice, text, image, etc.) of all spread spectrum communications transmitted.

(g) The transmitter power must not exceed 100 W.

§97.313 Transmitter power standards.

(a) An amateur station must use the minimum transmitter power necessary to carry out the desired communications.

(b) No station may transmit with a transmitter power exceeding 1.5 kW PEP.

(c) No station may transmit with a transmitter power exceeding 200 W PEP on:

(1) The 3.675-3.725 MHz, 7.10-7.15 MHz, 10.10-10.15 MHz and 21.1-21.2 MHz segments;

(2) The 28.1-28.5 MHz segment when the control operator is a Novice or Technician operator; or

(3) The 7.050-7.075 MHz segment when the station is within ITU Regions 1 or 3.

(d) No station may transmit with a transmitter power exceeding 25 W PEP on the VHF 1.25 m band when the control operator is a Novice operator.

(e) No station may transmit with a transmitter power exceeding 5 W PEP on the UHF 23 cm band when the control operator is a Novice operator.

(f) No station may transmit with a transmitter power exceeding 50 W PEP on the UHF 70 cm band from an area specified in footnote US7 to §2.106 of the FCC Rules, unless expressly authorized by the FCC after mutual agreement, on a case-by-case basis, between the EIC of the applicable field facility and the military area frequency coordinator at the applicable military base. An Earth station or telecommand station, however, may transmit on the 435-438 MHz segment with a maximum of 611 W effective radiated power (1 kW equivalent isotropically radiated power) without the authorization otherwise required. The transmitting antenna elevation angle between the lower half-power (–3 dB relative to the peak or antenna bore sight) point and the horizon must always be greater than 10°.

(g) No station may transmit with a transmitter power exceeding 50 W PEP on the 33 cm band from within 241 km of the boundaries of the White Sands Missile Range. Its boundaries are those portions of Texas and New Mexico bounded on the south by latitude 31° 41' North, on the east by longitude 104° 11' West, on the north by latitude 34° 30' North, and on the west by longitude 107° 30' West.

(h) No station may transmit with a transmitter power exceeding 50 W PEP on the 219-220 MHz segment of the 1.25 m band.

§97.315 Type acceptance of external RF power amplifiers.

(a) No more than 1 unit of 1 model of an external RF power amplifier capable of operation below 144 MHz may be constructed or modified during any calendar year by an amateur operator for use at a station without a grant of type acceptance. No amplifier capable of operation below 144 MHz may be constructed or modified by a non-amateur operator without a grant of type acceptance from the FCC.

(b) Any external RF power amplifier or external RF power amplifier kit (see §2.815 of the FCC Rules), manufactured, imported or modified for use in a station or attached at any station must be type accepted for use in the amateur service in accordance with Subpart J of Part 2 of the FCC Rules. This requirement does not apply if one or more of the following conditions are met:

(1) The amplifier is not capable of operation on frequencies below 144 MHz. For the purpose of this part, an amplifier will be deemed to be incapable of operation below 144 MHz if it is not capable of being easily modified to increase its amplification characteristics below 120 MHz and either:

(i) The mean output power of the amplifier decreases, as frequency decreases from 144 MHz, to a point where 0 dB or less gain is exhibited at 120 MHz; or

(ii) The amplifier is not capable of amplifying signals below 120 MHz even for brief periods without sustaining permanent damage to its amplification circuitry.

(2) The amplifier was manufactured before April 28, 1978, and has been issued a marketing waiver by the FCC, or the amplifier was purchased

before April 28, 1978, by an amateur operator for use at that amateur operator's station.

(3) The amplifier was:

  (i) Constructed by the licensee, not from an external RF power amplifier kit, for use at the licensee's station; or

  (ii) Modified by the licensee for use at the licensee's station.

(4) The amplifier is sold by an amateur operator to another amateur operator or to a dealer.

(5) The amplifier is purchased in used condition by an equipment dealer from an amateur operator and the amplifier is further sold to another amateur operator for use at that operator's station.

(c) A list of type accepted equipment may be inspected at FCC headquarters in Washington, DC or at any FCC field location. Any external RF power amplifier appearing on this list as type accepted for use in the amateur service may be marketed for use in the amateur service.

§97.317 Standards for type acceptance of external RF power amplifiers.

(a) To receive a grant of type acceptance, the amplifier must satisfy the spurious emission standards of §97.307(d) or (e) of this Part, as applicable, when the amplifier is:

(1) Operated at its full output power;

(2) Placed in the "standby" or "off" positions, but still connected to the transmitter; and

(3) Driven with at least 50 W mean RF input power (unless higher drive level is specified).

(b) To receive a grant of type acceptance, the amplifier must not be capable of operation on any frequency or frequencies between 24 MHz and 35 MHz. The amplifier will be deemed incapable of such operation if it:

(1) Exhibits no more than 6 dB gain between 24 MHz and 26 MHz and between 28 MHz and 35 MHz. (This gain will be determined by the ratio of the input RF driving signal (mean power measurement) to the mean RF output power of the amplifier); and

(2) Exhibits no amplification (0 dB gain) between 26 MHz and 28 MHz.

(c) Type acceptance may be denied when denial would prevent the use of these amplifiers in services other than the amateur service. The following features will result in dismissal or denial of an application for the type acceptance:

(1) Any accessible wiring which, when altered, would permit operation of the amplifier in a manner contrary to the FCC Rules;

(2) Circuit boards or similar circuitry to facilitate the addition of components to change the amplifier's operating characteristics in a manner contrary to the FCC Rules;

(3) Instructions for operation or modification of the amplifier in a manner contrary to the FCC Rules;

(4) Any internal or external controls or adjustments to facilitate operation of the amplifier in a manner contrary to the FCC Rules;

(5) Any internal RF sensing circuitry or any external switch, the purpose of which is to place the amplifier in the transmit mode;

(6) The incorporation of more gain in the amplifier than is necessary to operate in the amateur service; for purposes of this paragraph, the amplifier must:
  (i) Not be capable of achieving designed output power when driven with less than 40 W mean RF input power;
  (ii) Not be capable of amplifying the input RF driving signal by more than 15 dB, unless the amplifier has a designed transmitter power of less than 1.5 kW (in such a case, gain must be reduced by the same number of dB as the transmitter power relationship to 1.5 kW; This gain limitation is determined by the ratio of the input RF driving signal to the RF output power of the amplifier where both signals are expressed in peak envelope power or mean power);
  (iii) Not exhibit more gain than permitted by paragraph (c)(6)(ii) of this Section when driven by an RF input signal of less than 50 W mean power; and
  (iv) Be capable of sustained operation at its designed power level.
(7) Any attenuation in the input of the amplifier which, when removed or modified, would permit the amplifier to function at its designed transmitter power when driven by an RF frequency input signal of less than 50 W mean power; or
(8) Any other features designed to facilitate operation in a telecommunication service other than the Amateur Radio Services, such as the Citizens Band (CB) Radio Service.

Subpart E—Providing Emergency Communications

§97.401 Operation during a disaster.

(a) When normal communication systems are overloaded, damaged or disrupted because a disaster has occurred, or is likely to occur, in an area where the amateur service is regulated by the FCC, an amateur station may make transmissions necessary to meet essential communication needs and facilitate relief actions.
(b) When normal communication systems are overloaded, damaged or disrupted because a natural disaster has occurred, or is likely to occur, in an area where the amateur service is not regulated by the FCC, a station assisting in meeting essential communication needs and facilitating relief actions may do so only in accord with ITU Resolution No. 640 (Geneva, 1979). The 80 m, 75 m, 40 m, 30 m, 20 m, 17 m, 15 m, 12 m, and 2 m bands may be used for these purposes.
(c) When a disaster disrupts normal communication systems in a particular area, the FCC may declare a temporary state of communication emergency. The declaration will set forth any special conditions and special rules to be observed by stations during the communication emergency. A request for a declaration of a temporary state of emergency should be directed to the EIC in the area concerned.
(d) A station in, or within 92.6 km of, Alaska may transmit emissions J3E and R3E on the channel at 5.1675 MHz for emergency communications. The channel must be shared with stations licensed in the Alaska-private fixed service. The transmitter power must not exceed 150 W.

§97.403 Safety of life and protection of property.

No provision of these rules prevents the use by an amateur station of any means of radiocommunication at its disposal to provide essential communication needs in connection with the immediate safety of human life and immediate protection of property when normal communication systems are not available.

§97.405 Station in distress.

(a) No provision of these rules prevents the use by an amateur station in distress of any means at its disposal to attract attention, make known its condition and location, and obtain assistance.

(b) No provision of these rules prevents the use by a station, in the exceptional circumstances described in paragraph (a), of any means of radiocommunications at its disposal to assist a station in distress.

§97.407 Radio amateur civil emergency service.

(a) No station may transmit in RACES unless it is an FCC-licensed primary, club, or military recreation station and it is certified by a civil defense organization as registered with that organization, or it is an FCC-licensed RACES station. No person may be the control operator of a RACES station, or may be the control operator of an amateur station transmitting in RACES unless that person holds a FCC-issued amateur operator license and is certified by a civil defense organization as enrolled in that organization.

(b) The frequency bands and segments and emissions authorized to the control operator are available to stations transmitting communications in RACES on a shared basis with the amateur service. In the event of an emergency which necessitates the invoking of the President's War Emergency Powers under the provisions of §706 of the Communications Act of 1934, as amended, 47 U.S.C. §606, RACES stations and amateur stations participating in RACES may only transmit on the following frequencies:

(1) The 1800-1825 kHz, 1975-2000 kHz, 3.50-3.55 MHz, 3.93-3.98 MHz, 3.984-4.000 MHz, 7.079-7.125 MHz, 7.245-7.255 MHz, 10.10-10.15 MHz, 14.047-14.053 MHz, 14.22-14.23 MHz, 14.331-14.350 MHz, 21.047-21.053 MHz, 21.228-21.267 MHz, 28.55-28.75 MHz, 29.237-29.273 MHz, 29.45-29.65 MHz, 50.35-50.75 MHz, 52-54 MHz, 144.50-145.71 MHz, 146-148 MHz, 2390-2450 MHz segments;

(2) The 1.25 m, 70 cm and 23 cm bands; and

(3) The channels at 3.997 MHz and 53.30 MHz may be used in emergency areas when required to make initial contact with a military unit and for communications with military stations on matters requiring coordination.

(c) A RACES station may only communicate with:

(1) Another RACES station;

(2) An amateur station registered with a civil defense organization;

(3) A United States Government station authorized by the responsible agency to communicate with RACES stations;

(4) A station in a service regulated by the FCC whenever such communication is authorized by the FCC.

(d) An amateur station registered with a civil defense organization may only communicate with:

(1) A RACES station licensed to the civil defense organization with which the amateur station is registered;

(2) The following stations upon authorization of the responsible civil defense official for the organization with which the amateur station is registered:

    (i) A RACES station licensed to another civil defense organization;

    (ii) An amateur station registered with the same or another civil defense organization;

    (iii) A United States Government station authorized by the responsible agency to communicate with RACES stations; and

    (iv) A station in a service regulated by the FCC whenever such communication is authorized by the FCC.

(e) All communications transmitted in RACES must be specifically authorized by the civil defense organization for the area served. Only civil defense communications of the following types may be transmitted:

(1) Messages concerning impending or actual conditions jeopardizing the public safety, or affecting the national defense or security during periods of local, regional, or national civil emergencies;

(2) Messages directly concerning the immediate safety of life of individuals, the immediate protection of property, maintenance of law and order, alleviation of human suffering and need, and the combating of armed attack or sabotage;

(3) Messages directly concerning the accumulation and dissemination of public information or instructions to the civilian population essential to the activities of the civil defense organization or other authorized governmental or relief agencies; and

(4) Communications for RACES training drills and tests necessary to ensure the establishment and maintenance of orderly and efficient operation of the RACES as ordered by the responsible civil defense organizations served. Such drills and tests may not exceed a total time of 1 hour per week. With the approval of the chief officer for emergency planning the applicable State, Commonwealth, District or territory, however, such tests and drills may be conducted for a period not to exceed 72 hours no more than twice in any calendar year.

Subpart F—Qualifying Examinations Systems

§97.501 Qualifying for an amateur operator license.

Each applicant for the grant of a new amateur operator license or for the grant of a modified license to show a higher operator class, must pass or otherwise receive credit for the examination elements specified for the class of operator license sought:

    (a) Amateur Extra Class operator: Elements 1(C), 2, 3(A), 3(B), 4(A) and 4(B);

    (b) Advanced Class operator: Elements 1(B) or 1(C), 2, 3(A), 3(B) and 4(A);

    (c) General Class operator: Elements 1(B) or 1(C), 2, 3(A) and 3(B);

    (d) Technician Plus Class operator: Elements 1(A) or 1(B) or 1(C), 2, and 3(A).

    (e) Technician Class operator: Elements 2 and 3(A).

    (f) Novice Class operator: Elements 1(A) or 1(B) or 1(C), and 2.

§97.503 Element standards.

(a) A telegraphy examination must be sufficient to prove that the examinee has the ability to send correctly by hand and to receive correctly by ear texts in the international Morse code at not less than the prescribed speed, using all the letters of the alphabet, numerals 0-9, period, comma, question mark, slant mark and prosigns AR, BT and SK.

(1) Element 1(A): 5 words per minute;

(2) Element 1(B): 13 words per minute;

(3) Element 1(C): 20 words per minute.

(b) A written examination must be such as to prove that the examinee possesses the operational and technical qualifications required to perform properly the duties of an amateur service licensee. Each written examination must be comprised of a question set as follows:

(1) Element 2: 30 questions concerning the privileges of a Novice Class operator license. The minimum passing score is 22 questions answered correctly.

(2) Element 3(A): 25 questions concerning the additional privileges of a Technician Class operator license. The minimum passing score is 19 questions answered correctly.

(3) Element 3(B): 25 questions concerning the additional privileges of a General Class operator license. The minimum passing score is 19 questions answered correctly.

(4) Element 4(A): 50 questions concerning the additional privileges of an Advanced Class operator license. The minimum passing score is 37 questions answered correctly.

(5) Element 4(B): 40 questions concerning the additional privileges of an Amateur Extra Class operator license. The minimum passing score is 30 questions answered correctly.

(c) The topics and number of questions required in each question set are listed below for the appropriate examination element:

Topics	Element: 2	3(A)	3(B)	4(A)	4(B)
(1) FCC Rules for the amateur radio services	10	5	4	6	8
(2) Amateur station operating procedures	2	3	3	1	4
(3) Radio wave propagation characteristics of amateur service frequency bands	1	3	3	2	2
(4) Amateur radio practices	4	4	5	4	4
(5) Electrical principles as applied to amateur station equipment	4	2	2	10	6
(6) Amateur station equipment circuit components	2	2	1	6	4
(7) Practical circuits employed in amateur station equipment	2	1	1	10	4
(8) Signals and emissions transmitted by amateur stations	2	2	2	6	4
(9) Amateur station antennas and feed lines	3	3	4	5	4

§97.505 Element credit.

(a) The administering VEs must give credit as specified below to an examinee holding any of the following documents:

(1) An unexpired (or expired but within the grace period for renewal) FCC-granted Advanced Class operator license document: Elements 1(B), 2, 3(A), 3(B), and 4(A).

(2) An unexpired (or expired but within the grace period for renewal) FCC-granted General Class operator license document: Elements 1(B), 2, 3(A), and 3(B).

(3) An unexpired (or expired but within the grace period for renewal) FCC-granted Technician Plus Class operator (including a Technician Class operator license granted before February 14, 1991) license document: Elements 1(A), 2, and 3(A).

(4) An unexpired (or expired but within the grace period for renewal) FCC-granted Technician Class operator license document: Elements 2 and 3(A).

(5) An unexpired (or expired but within the grace period for renewal) FCC-granted Novice Class operator license document: Elements 1(A) and 2.

(6) A CSCE: Each element the CSCE indicates the examinee passed within the previous 365 days.

(7) An unexpired (or expired for less than 5 years) FCC-issued commercial radiotelegraph operator license document or permit: Element 1(C).

(8) An expired or unexpired FCC-issued Technician Class operator license document granted before March 21, 1987: Element 3(B).

(9) An expired or unexpired FCC-issued Technician Class license document granted before February 14, 1991: Element 1(A).

(10) An unexpired (or expired but within the grace period for renewal), FCC-granted Novice, Technician Plus (including a Technician Class operator license granted before February 14, 1991), General, or Advanced Class operator license document, and a FCC Form 610 containing:

   (i) A physician's certification stating that because the person is an individual with a severe handicap, the duration of which will extend for more than 365 days beyond the date of the certification, the person is unable to pass a 13 or 20 words per minute telegraphy examination; and

   (ii) A release signed by the person permitting the disclosure to the FCC of medical information pertaining to the person's handicap: Element 1(C).

(b) No examination credit, except as herein provided, shall be allowed on the basis of holding or having held any other license grant or document.

§97.507 Preparing an examination.

(a) Each telegraphy message and each written question set administered to an examinee must be prepared by a VE who has been granted an Amateur Extra Class operator license. A telegraphy message or written question set, however, may also be prepared for the following elements by a VE who has been granted an FCC operator license of the class indicated:

(1) Element 3(B): Advanced Class operator.

(2) Elements 1(A) and 3(A): Advanced or General Class operator.

(3) Element 2: Advanced, General, Technician, or Technician Plus Class operator.

(b) Each question set administered to an examinee must utilize questions taken from the applicable question pool.

(c) Each telegraphy message and each written question set administered to an examinee for an amateur operator license must be prepared, or obtained from a supplier, by the administering VEs according to instructions from the coordinating VEC.

(d) A telegraphy examination must consist of a message sent in the international Morse code at no less than the prescribed speed for a minimum of 5 minutes. The message must contain each required telegraphy character at least once. No message known to the examinee may be administered in a telegraphy examination. Each 5 letters of the alphabet must be counted as 1 word. Each numeral, punctuation mark and prosign must be counted as 2 letters of the alphabet.

§97.509 Administering VE requirements.

(a) Each examination for an amateur operator license must be administered by 3 administering VEs at an examination session coordinated by a VEC. Before the session, the administering VEs must make a public announcement stating the location and time of the session. The number of examinees at the session may be limited.

(b) Each administering VE must:

(1) Be accredited by the coordinating VEC;

(2) Be at least 18 years of age;

(3) Be a person who has been granted an FCC amateur operator license document of the class specified below:

(i) Amateur Extra, Advanced, or General Class in order to administer a Novice, Technician, or Technician Plus Class operator license examination;

(ii) Amateur Extra Class in order to administer a General, Advanced, or Amateur Extra Class operator license examination.

(4) Not be a person whose grant of an amateur station license or amateur operator license has ever been revoked or suspended.

(c) Each administering VE must be present and observing the examinee throughout the entire examination. The administering VEs are responsible for the proper conduct and necessary supervision of each examination. The administering VEs must immediately terminate the examination upon failure of the examinee to comply with their instructions.

(d) No VE may administer an examination to his or her spouse, children, grandchildren, stepchildren, parents, grandparents, stepparents, brothers, sisters, stepbrothers, stepsisters, aunts, uncles, nieces, nephews, and in-laws.

(e) No VE may administer or certify any examination by fraudulent means or for monetary or other consideration including reimbursement in any amount in excess of that permitted. Violation of this provision may result in the revocation of the grant of the VE's amateur station license and the suspension of the grant of the VE's amateur operator license.

(f) No examination that has been compromised shall be administered to any examinee. Neither the same telegraphy message nor the same question set may be re-administered to the same examinee.

(g) Passing a telegraphy receiving examination is adequate proof of an examinee's ability to both send and receive telegraphy. The administering VEs, however, may also include a sending segment in a telegraphy examination.

(h) Upon completion of each examination element, the administering VEs must immediately grade the examinee's answers. The administering VEs are responsible for determining the correctness of the examinee's answers.

(i) When the examinee is credited for all examination elements required for the operator license sought, the administering VEs must certify on the examinee's application document that the applicant is qualified for the license.

(j) When the examinee does not score a passing grade on an examination element, the administering VEs must return the application document to the examinee and inform the examinee of the grade.

(k) The administering VEs must accommodate an examinee whose physical disabilities require a special examination procedure. The administering VEs may require a physician's certification indicating the nature of the disability before determining which, if any, special procedures must be used.

(l) The administering VEs must issue a CSCE to an examinee who scores a passing grade on an examination element.

(m) Within 10 days of the administration of a successful examination for an amateur operator license, the administering VEs must submit the application document to the coordinating VEC.

§97.511 Examinee conduct.

Each examinee must comply with the instructions given by the administering VEs.

§97.513 [Removed and Reserved]

§97.515 [Reserved]

§97.517 [Reserved]

§97.519 Coordinating examination sessions.

(a) A VEC must coordinate the efforts of VEs in preparing and administering examinations.

(b) At the completion of each examination session, the coordinating VEC must collect the FCC Forms 610 documents and test results from the administering VEs. Within 10 days of collecting the FCC Forms 610 documents, the coordinating VEC must screen and, for qualified examinees, forward electronically or on diskette the data contained on the FCC Forms 610 documents, or forward the FCC Form 610 documents to: FCC, 1270 Fairfield Road, Gettysburg, PA 17325-7245. When the data is forwarded electronically, the coordinating VEC must retain the FCC Forms 610 documents for at least fifteen months and make them available to the FCC upon request.

(c) Each VEC must make any examination records available to the FCC, upon request.

(d) The FCC may:

(1) Administer any examination element itself;

(2) Readminister any examination element previously administered by VEs, either itself or under the supervision of a VEC or VEs designated by the FCC; or

(3) Cancel the operator/primary station license of any licensee who fails to appear for readministration of an examination when directed by the FCC, or who does not successfully complete any required element that is readministered. In an instance of such cancellation, the person will be granted an operator/primary station license consistent with completed examination elements that have not been invalidated by not appearing for, or by failing, the examination upon readministration.

§97.521 VEC qualifications.

No organization may serve as a VEC unless it has entered into a written agreement with the FCC. The VEC must abide by the terms of the agreement. In order to be eligible to be a VEC, the entity must:

(a) Be an organization that exists for the purpose of furthering the amateur service;

(b) Be capable of serving as a VEC in at least the VEC region (see Appendix 2) proposed;

(c) Agree to coordinate examinations for any class of amateur operator license;

(d) Agree to assure that, for any examination, every examinee qualified under these rules is registered without regard to race, sex, religion, national origin or membership (or lack thereof) in any amateur service organization.

§97.523 Question pools.

All VECs must cooperate in maintaining one question pool for each written examination element. Each question pool must contain at least 10 times the number of questions required for a single examination. Each question pool must be published and made available to the public prior to its use for making a question set. Each question on each VEC question pool must be prepared by a VE holding the required FCC-issued operator license. See §97.507(a) of this Part.

§97.525 Accrediting VEs.

(a) No VEC may accredit a person as a VE if:

(1) The person does not meet minimum VE statutory qualifications or minimum qualifications as prescribed by this Part;

(2) The FCC does not accept the voluntary and uncompensated services of the person;

(3) The VEC determines that the person is not competent to perform the VE functions; or

(4) The VEC determines that questions of the person's integrity or honesty could compromise the examinations.

(b) Each VEC must seek a broad representation of amateur operators to be VEs. No VEC may discriminate in accrediting VEs on the basis of race, sex, religion or national origin; nor on the basis of membership (or lack thereof) in an amateur service organization; nor on the basis of the person accepting or declining to accept reimbursement.

§97.527 Reimbursement for expenses.

(a) VEs and VECs may be reimbursed by examinees for out-of-pocket expenses incurred in preparing, processing, administering, or coordinating an examination for an amateur operator license.

(b) The maximum amount of reimbursement from any one examinee for any one examination at a particular session regardless of the number of examination elements taken must not exceed that announced by the FCC in a Public Notice. (The basis for the maximum fee is $4.00 for 1984, adjusted annually each January 1 thereafter for changes in the Department of Labor Consumer Price Index.)

Appendix 1—Places Where the Amateur Service is Regulated by the FCC

In ITU Region 2, the amateur service is regulated by the FCC within the territorial limits of the 50 United States, District of Columbia, Caribbean Insular areas [Commonwealth of Puerto Rico, United States Virgin Islands (50 islets and cays) and Navassa Island], and Johnston Island (Islets East, Johnston, North and Sand) and Midway Island (Islets Eastern and Sand) in the Pacific Insular areas.

In ITU Region 3, the amateur service is regulated by the FCC within the Pacific Insular territorial limits of American Samoa (seven islands), Baker Island, Commonwealth of Northern Mariannas Islands, Guam Island, Howland Island, Jarvis Island, Kingman Reef, Kure Island, Palmyra Island (more than 50 islets) and Wake Island (Islets Peale, Wake and Wilkes).

Appendix 2—VEC Regions

1. Connecticut, Maine, Massachusetts, New Hampshire, Rhode Island and Vermont.
2. New Jersey and New York.
3. Delaware, District of Columbia, Maryland and Pennsylvania.
4. Alabama, Florida, Georgia, Kentucky, North Carolina, South Carolina, Tennessee and Virginia.
5. Arkansas, Louisiana, Mississippi, New Mexico, Oklahoma and Texas.
6. California.
7. Arizona, Idaho, Montana, Nevada, Oregon, Utah, Washington and Wyoming.
8. Michigan, Ohio and West Virginia.
9. Illinois, Indiana and Wisconsin.
10. Colorado, Iowa, Kansas, Minnesota, Missouri, Nebraska, North Dakota and South Dakota.
11. Alaska.
12. Caribbean Insular areas.
13. Hawaii and Pacific Insular areas.

APPENDICES

Appendices List

Appendix 1
Extracts from the Communications Act of 1934

The complete text of the Communications Act of 1934 occupies a volume larger than this book. Only the parts most applicable to Amateur Radio are included here. The entire text can be ordered from the Government Printing Office. You can also find it in your nearest Government Document Depository Library; check with your local library.

Section 4(f)

(4)(A) The Commission, for purposes of preparing any examination for an amateur station operator license, may accept and employ the voluntary and uncompensated services of any individual who holds an amateur station operator license of a higher class than the class license for which the examination is being prepared. In the case of examinations for the highest class of amateur station operator license, the Commission may accept and employ such services of any individual who holds such class of license.

(B)(i) The Commission, for purposes of monitoring violations of any provision of this Act (and of any regulation prescribed by the Commission under this Act) relating to the Amateur Radio Service, may—

(I) recruit and train any individual licensed by the Commission to operate an amateur station; and

(II) accept and employ the voluntary and uncompensated services of such individual.

(ii) The Commission, for purposes of recruiting and training individuals under clause (i) and for purposes of screening, annotating, and summarizing violation reports referred under clause (i), may accept and employ the voluntary and uncompensated services of any amateur station operator organization.

(iii) The functions of individuals recruited and trained under this subparagraph shall be limited to—

(I) the detection of improper Amateur Radio transmissions;

(II) the conveyance to Commission personnel of information which is essential to the enforcement of this Act (or regulations prescribed by the Commission under this Act) relating to the Amateur Radio Service; and

(III) issuing advisory notices, under the general direction of the Commission, to persons who apparently have violated any provision of this Act (or regulations prescribed by the Commission under this Act) relating to the Amateur Radio Service.

Nothing in this clause shall be construed to grant individuals recruited and trained under this subparagraph any authority to issue sanctions to violators or to take any enforcement action other than any action which the Commission may prescribe by rule.

(F) Any person who provides services under this paragraph shall not be considered, by reason of having provided such services, a Federal employee.

(G) The Commission, in accepting and employing services of individuals under subparagraphs (A), (B), and (C), shall seek to achieve a broad representation of individuals and organizations interested in amateur station operation.

(H) The Commission may establish rules of conduct and other regulations governing the service of individuals under this paragraph.

(I) With respect to the acceptance of voluntary uncompensated services for the preparation, processing, or administration of examinations for amateur station operator licenses pursuant to subparagraph (A) of this paragraph, individuals, or organizations which provide or coordinate such authorized volunteer services may recover from examinees reimbursement for out-of-pocket costs. The total amount of allowable cost reimbursement per examinee shall not exceed $4, adjusted annually every January 1 for changes in the Department of Labor Consumer Price Index.

* * * * * * * * * * * * * * *

Section 302

(a) The Commission may, consistent with the public interest, convenience, and necessity, make reasonable regulations (1) governing the interference potential of devices which in their operation are capable of emitting radio frequency energy...in sufficient degree to cause harmful interference to radio communications; and (2) establishing minimum performance standards for home electronic equipment and systems to reduce their susceptibility to interference from radio frequency energy. Such regulations shall be applicable to the manufacture, import, sale, offer for sale, or shipment of such devices and home electronic equipment and systems, and to the use of such devices.

(b) No person shall manufacture, import, sell, offer for sale, or ship devices or home electronic equipment and systems, or use devices, which fail to comply with regulations promulgated pursuant to this section.

* * * * * * * * * * * * * * *

Section 303

Except as otherwise provided in this Act, the Commission from time to time, as public convenience, interest or necessity requires, shall—

---

f) Make such regulations not inconsistent with law as it may deem necessary to prevent interference between stations and to carry out the provisions of this Act: *Provided, however,* that changes in the frequencies, authorized power, or in the times of operation of any station, shall not be made without the consent of the station licensee unless, after a public hearing, the Commission shall determine that such changes will promote public convenience or interest or will serve public necessity, or the provisions of this Act will be more fully complied with;

g) Study new uses for radio, provide for experimental uses of frequencies, and generally encourage the larger and more effective use of radio in the public interest;

_____

j) Have authority to make general rules and regulations requiring stations to keep such records...as it may deem desirable;

_____

l)1) Have the authority to prescribe the qualifications of station operators...and to issue [licenses] to persons who are found to be qualified by the Commission and who otherwise are legally eligible for employment in the United States...

2) Not withstanding paragraph (1) of this subsection, an individual to whom a radio station is licensed under the provisions of this Act may be issued an operator's license to operate that station.

3) In addition to amateur operator licenses which the Commission may issue to aliens pursuant to paragraph (2) of this subsection, and notwithstanding Section 301 of this Act and paragraph (1) of this subsection, the Commission may issue authorizations, under such conditions and terms as it may prescribe, to permit an alien licensed by his government as an Amateur Radio operator to operate his Amateur Radio station licensed by his government in the United States, its possessions, and the Commonwealth of Puerto Rico provided there is in effect a multilateral or bilateral agreement, to which the United States and the alien's government are parties, for such operation on a reciprocal basis by United States Amateur Radio operators. Other provisions of this Act and of the Administrative Procedure Act shall not be applicable to any request or application for or modification, suspension or cancellation of any such authorization.

* * * * * * * * * * * * * *

Section 310.

_(a)_ The station license required under this Act shall not be granted to or held by any foreign government or the representative thereof.

_(c)_ In addition to amateur station licenses which the Commission may issue to aliens pursuant to this Act, the Commission may issue authorizations, under such conditions and terms as it may prescribe, to permit an alien licensed by his government as an Amateur Radio operator to operate his Amateur Radio station licensed by his government in the United States, its possessions, and the Commonwealth of Puerto Rico provided there is in effect a multilateral or bilateral agreement, to which the United States and the alien's government are parties, for such operation on a reciprocal basis by United States Amateur Radio operators. Other provisions of this Act and of the Administrative Procedure Act shall not be applicable to any request or application for or modification, suspension or cancellation of any such authorization.

* * * * * * * * * * * * * *

Section 705(a)...No person receiving, assisting in receiving, transmitting, or assisting in transmitting, any interstate or foreign communication by wire or radio shall divulge or publish the existence, contents, substance, purport, effect, or meaning thereof,...to any person other than the addressee, his agent or attorney…[or] in response to a subpena issued by a court of competent jurisdiction, or on demand of other lawful authority… No person not being entitled thereto shall receive or assist in receiving any interstate or foreign communication by radio and use such information (or any information therein contained) for his own benefit or for the benefit of another not entitled thereto…. This section shall not apply to the receiving, divulging, publishing, or utilizing the contents of any radio communication which is transmitted by any station for the use of the general public, which relates to ships, aircraft, vehicles, or persons in distress, or which is transmitted by an amateur radio station operator or by a citizens band radio operator.

Appendix 2
Extracts from Title 47
Code of Federal Regulations

The complete set of FCC Regulations, (Title 47 of the Code of Federal Regulations) occupies 5 volumes. The entire set is updated yearly, and can be ordered from the Government Printing Office. You may also find it in your nearest Government Document Depository Library; check with your local library.

§0.314 Additional authority delegated.

(x) When deemed necessary by the Engineer-in-Charge of a Commission field facility to assure compliance with the Rules, a station licensee shall maintain a record of such operating and maintenance records as may be necessary to resolve conditions of interference or deficient technical operation.

✳ ✳ ✳ ✳ ✳ ✳ ✳ ✳ ✳ ✳ ✳ ✳ ✳ ✳ ✳

Classification of Emissions

§2.201 Emission, modulation and transmission characteristics.

The following system of designating emission, modulation and transmission characteristics shall be employed.

(a) Emissions are designated according to their classification and their necessary bandwidth.

(b) A minimum of three symbols are used to describe the basic characteristics of radio waves. Emissions are classified and symbolized according to the following characteristics:

(1) First symbol—type of modulation of the main carrier;

(2) Second symbol—nature of signal(s) modulating the main carrier;

(3) Third symbol—type of information to be transmitted.

Note: A fourth and fifth symbol are provided for additional information and are shown in Appendix 6, Part A of the ITU Radio Regulations. Use of the fourth and fifth symbol is optional. Therefore, the symbols may be used as described in Appendix 6, but are not required by the Commission.

(c) First symbol—types of modulation of the main carrier:

(1) Emission of an unmodulated carrier .... N
(2) Emission in which the main carrier is amplitude-modulated (including cases where subcarriers are angle-modulated):
  —Double sideband .... A
  —Single sideband, full carrier .... H
  —Single sideband, reduced or variable level carrier .... R

—Single sideband, suppressed carrier .... J
—Independent sidebands .... B
—Vestigial sideband C
(3) Emission in which the main carrier is angle modulated:
—Frequency modulation.. F
—Phase modulation .... G
Note: Whenever frequency modulation (F) is indicated, phase modulation (G) is also acceptable.
(4) Emission in which the main carrier is amplitude and angle-modulated either simultaneously or in a pre-established sequence .... D
(5) Emission of pulses[1]
—Sequence of unmodulated pulses .... P
—A sequence of pulses:
—Modulated in amplitude .... K
—Modulated in width/duration L
—Modulated in position/phase .... M
—In which the carrier is angle-modulated during the period of the pulse .... Q
—Which is a combination of the foregoing or is produced by other means V
(6) Cases not covered above, in which an emission consists of the main carrier modulated, either simultaneously or in a pre-established sequence in a combination of two or more of the following modes: amplitude, angle, pulse... W
(7) Cases not otherwise covered X

(d) Second Symbol—nature of signal(s) modulating the main carrier:

(1) No modulating signal 0
(2) A single channel containing quantized or digital information without the use of a modulating subcarrier, excluding time-division multiplex 1
(3) A single channel containing quantized or digital information with the use of a modulating subcarrier, excluding time-division multiplex 2
(4) A single channel containing analog information 3
(5) Two or more channels containing quantized or digital information 7
(6) Two or more channels containing analog information.................................. 8
(7) Composite system with one or more channels containing quantized or digital information, together with one or more channels containing analog information 9
(8) Cases not otherwise covered X

(e) Third Symbol—type of information to be transmitted:[2]

(1) No information transmitted .... N
(2) Telegraphy, for aural reception .... A
(3) Telegraphy, for automatic reception... B
(4) Facsimile.. C

[1]Emissions where the main carrier is directly modulated by a signal which has been coded into quantized form (e.g., pulse code modulation) should be designated under (2) or (3)):
[2]In this context the word "information" does not include information of a constant, unvarying nature such as is provided by standard frequency emissions, continuous wave and pulse radars, etc.

(5) Data transmission, telemetry, telecommand .... D
(6) Telephony (including sound broadcasting) .... E
(7) Television (video) .... F
(8) Combination of the above .... W
(9) Cases not otherwise covered X

(f) Type *B* emission: As an exception to the above principles, damped waves are symbolized in the Commission's rules and regulations as type *B* emission. The use of type B emissions is forbidden.

(g) Whenever the full designation of an emission is necessary, the symbol for that emission, as given above, shall be preceded by the necessary bandwidth of the emission as indicated in §2.202(b)(1).

★ ★ ★ ★ ★ ★ ★ ★ ★ ★ ★ ★ ★ ★ ★

§2.202 Bandwidths.

(a) *Occupied bandwidth.* The frequency bandwidth such that, below its lower and above its upper frequency limits, the mean powers radiated are each equal to 0.5 percent of the total mean power radiated by a given emission. In some cases, for example multi-channel frequency-division systems, the percentage of 0.5 percent may lead to certain difficulties in the practical application of the definitions of occupied and necessary bandwidth; in such cases a different percentage may prove useful.

(b) *Necessary bandwidth.* For a given class of emission, the minimum value of the occupied bandwidth sufficient to ensure the transmission of information at the rate and with the quality required for the system employed, under specified conditions. Emissions useful for the good functioning of the receiving equipment as, for example, the emission corresponding to the carrier of reduced carrier systems, shall be included in the necessary bandwidth.

★ ★ ★ ★ ★ ★ ★ ★ ★ ★ ★ ★ ★ ★ ★

§2.815 External radio frequency power amplifiers.

(a) As used in this Part, an external radio frequency power amplifier is any device which, (1) when used in conjunction with a radio transmitter as a signal source is capable of amplification of that signal, and (2) is not an integral part of a radio transmitter as manufactured.

(b) After April 27, 1978, no person shall manufacture, sell or lease, offer for sale or lease (including advertising for sale or lease), or import, ship, or distribute for the purpose of selling or leasing or offering for sale or lease, any external radio frequency power amplifier or amplifier kit capable of operation on any frequency or frequencies between 24 and 35 MHz.

NOTE: For purposes of this part, the amplifier will be deemed incapable of operation between 24 and 35 MHz if:

(1) The amplifier has no more than 6 decibels of gain between 24 and 26 MHz and between 28 and 35 MHz. (This gain is determined by the ratio of the input RF driving signal (mean power measurement) to the mean RF output power of the amplifier.); and

(2) The amplifier exhibits no amplification (0 decibels of gain) between 26 and 28 MHz.

(c) No person shall manufacture, sell or lease, offer for sale or lease (including advertising for sale or lease) or import, ship or distribute for the purpose of selling or leasing or offering for sale or lease, any external radio frequency power amplifier or amplifier kit capable of operation on any frequency or frequencies below 144 MHz unless the amplifier has received a grant of type acceptance in accordance with Subpart J of this part and Subpart C of Part 97 or other relevant parts of this chapter. No more than 10 external radio frequency power amplifiers or amplifier kits may be constructed for evaluation purposes in preparation for the submission of an application for a grant of type acceptance.

NOTE: For the purposes of this part, an amplifier will be deemed incapable of operation below 144 MHz if the amplifier is not capable of being easily modified to increase its amplification characteristics below 120 MHz, and either:

(1) The mean output power of the amplifier decreases, as frequency decreases from 144 MHz, to a point where 0 decibels or less gain is exhibited at 120 MHz and below 120 MHz, or

(2) The amplifier is not capable of even short periods of operation below 120 MHz without sustaining permanent damage to its amplification circuitry.

(d) The proscription in paragraph (b) of this section shall not apply to the marketing, as defined in that paragraph, by a licensed amateur radio operator to another licensed amateur radio operator of an external radio frequency power amplifier fabricated in not more than one unit of the same model in a calendar year by that operator provided the amplifier is for the amateur operator's personal use at his licensed amateur radio station and the requirements of §§97.315 and 97.317 of this chapter are met.

(e) The proscription in paragraph (c) of this section shall not apply in the marketing, as defined in that paragraph, by a licensed amateur radio operator to another licensed amateur radio operator of an external radio frequency power amplifier if the amplifier is for the amateur operator's personal use at his licensed amateur radio station and the requirements of §§97.315 and 97.317 of this chapter are met.

Appendix 3
Report and Orders Affecting Part 97 Since 1989

Where no effective date is shown, order was effective 30 days after publication in the *Federal Register*.

DATES: RELEASED	EFFECTIVE	DOCKET#	SUBJECT
Jun 9,'89	Sep 1,'89	88-139	**Reorganization and Deregulation of Part 97 of the Rules Governing the Amateur Radio Service.** Complete rewrite.
Sep 7,'89	—	88-139	ERRATA—corrects index, §§97.5, 97.15, 97.109, 97.119, 97.207, 97.209.
May 16,'90	Jul 2,'90		**Waiver of Parts 2 and 97 of the Rules Concerning Frequency Sharing Requirements Applicable to the Amateur service in Portions of Colorado and Wyoming.** Not a Report & Order—suspends but does not change parts of §97.303(g).
Jul 20,'90	Sep 10,'90	88-139	Minor & technical amendments & clarifications to §§97.19, 97.21, 97.119, 97.301, 97.305.
Dec 27,'90	Feb 14,'91	90-55	**Amendment of Part 97 of the Commission's Rules Concerning the Establishment of a Codeless Class of Amateur Operator License.** §§97.119, 97.301, 97.501.
Jan 2, '91	Feb 14, '91	90-356	**Amendment of the Amateur Radio Service Rules to Make the Sevice More Accessible to Persons with Handicaps.** §§97.3. 97.505, 97.511.
Jan 18,'91	Mar 16,'91	90-100	**Amendment of the Amateur Service Rules to Relocate the Novice and Technician Operator Class frequency Segment within the Amateur Service 80 Meter Band.** §§97.301, 97.313.
Apr 17,'91	May 29,'91	89-552	**Amendment of Part 90 of the**

DATES: RELEASED	EFFECTIVE	DOCKET#	SUBJECT
			Amateur Service. §97.21-deletes station location requirement.
Dec 2,'93	Feb 1,'94	92-289	**Amendment of the Amateur Service Rules Concerning the 222-225 MHz and 1240-1300 MHz Frequency Bands.** 222-222.15 subband created. Novices not granted repeater control op. privileges on 1.25 m or 23 cm bands. §§97.201, 97.205, 97.301.
Dec 29,'93	—		**Amendment of the Amateur Service Rules to Establish Station Call Sign Administrators for Club and Military Recreation Stations.** Cancellation of program concurrent with vanity NPRM
Apr 13,'94	Jun 1,'94	93-85	**Amendment of Part 97 of the Commission's Rules Concerning Message Forwarding Systems in the Amateur Service.** §§97.3, 97.109, 97.205, 97.219.
Oct 24,'94	Dec 20,'94		**Amendment of the Amateur Service Rules to Change Procedures for Filing an Amateur Service License Application and to Make Other Procedural Changes.** Creates Tech Plus class; allows electronic filing, Form 610-R; adds requirement to change address; drops requirement for license document in hand. §§97.5, 97.7, 97.9, 97.17, 97.21, 97.23, 97.25, 97.27, 97.29, 97.301, 97.501, 97.505, 97.507, 97.509, 97.511, 97.519.
Feb. 1,'95	Mar. 24,'95	93-305	**Amendment of the Amateur Service Rules to Implement a Vanity Call Sign System.** §§97.3, 97.17, 97.19, 97.21.
Feb 6,'95	—	93-61	**Amendment of Part 90 of the Commission's Rules to Adopt Regulations for Automatic Vehicle Monitoring Systems.** Does not modify Part 97, but changes AVMS to Location and Monitoring Service (LMS) and expands its subbands to cover the entire amateur band at 902-928 MHz. We are still secondary to the LMS.
Feb 17,'95	—	94-32	**Allocation of Spectrum Below 5 GHz Transferred from Federal Government Use.** This does not modify Part 97, but does

modify §2.106 and Part 15 to, among other things, give amateurs primary status in the 2390-2400 and 2402-2417 MHz bands.

DATES:		DOCKET#	SUBJECT
Mar 17,'95	Apr 26,'95	93-40	**Allocation of the 219-220 MHz Band for Use by the Amateur Radio Service.** §§2.106, 97.201, 97.301, 97.303, 97.305, 97.307, 97.313.
Apr 27,'95	Jul 1,'95	94-59	**Amendment of Part 97 of the Commission's Rules Concerning HF Digital Communications in the Amateur Service.** §§97.109, 97.221
Sep 20, '95	Oct 31, '95	—	**Amendment of Parts 80, 90 and 97 of the Commission's Rules to reflect Bureau name changes and to make other editorial changes.** Changes "Private Radio" to "Wireless Telecommunications" §§97.15, 97.19, 97.207
Oct 2, '95	Nov 17, '95	93-305	**Amendment of the Amateur Service Rules to Implement a Vanity Call Sign System.** Limits requests for call signs reflecting Region 11, 12, &13. Requires applicant seeking deceased relative's call to have appropriate class of license. Fixes a renewal problem. §§97.17, 97.19, & 97.21.
Oct 11, '95	Nov 1, '95	—	**Amendment of the Amateur Service Rules to Clarify Use of CLOVER, G-TOR, and PacTOR Digital Codes.** §97.309.
Nov 30, '95	Mar 7, '96	95-473	**In the Matter of Streamlining the Commission's Antenna Structure Clearance Procedure and Revision of Part 17 of the Commission's Rules Concerning Construction, Marking, and Lighting of Antenna Structures.** Reflects changes to Part 17. §97.15(d).
Feb 28, '96	Apr 11, '96	96-74	**In the Matter of Amendment of Part 97 of the Commission's Rules to Conform the Amateur Service Rules to the Provisions of the Telecommunications Act of 1996.** Drops VE & VEC Conflict of interest and reimbursement rules. §§97.509, 97.521, & 97.527.
Mar 22, '96	—	93-40	**Allocation of the 219-220 MHz Band for Use by the Amateur Radio Service.** Minor amendment of §97.303.

Appendix 4
Reciprocal Operating Agreement with Canada

Convention Between Canada and the United States of America, Relating to the operation by Citizens of Either Country of Certain Radio Equipment or Stations in the Other Country (Effective May 15, 1952)

Article III

It is agreed that persons holding appropriate amateur licenses issued by either country may operate their amateur stations in the territory of the other country under the following conditions:

(a) Each visiting amateur may be required to register and receive a permit before operating any amateur station licensed by his government.

(b) The visiting amateur will identify his station by:

(1) Radiotelegraph operation. The amateur call sign issued to him/her by the licensing country followed by a slant (/) sign and the amateur call sign prefix and call area number of the country he is visiting.

(2) Radiotelephone operation. The amateur call sign in English issued to him by the licensing country followed by the words, "fixed" "portable" or "mobile," as appropriate, and the amateur call sign prefix and call area number of the country he is visiting.

(c) Each amateur station shall indicate at least once during each contact with another station its geographical location as nearly as possible by city and state or city and province.

(d) In other respects the amateur station shall be operated in accordance with the laws and regulations of the country in which the station is temporarily located.

Appendix 5
Extracts from the International Radio Regulations

Article 1—Terms and Definitions

Section III. Radio Services

§3.34 Amateur Service: A radiocommunication service for the purpose of self-training, intercommunication and technical investigations carried out by amateurs, that is, by duly authorized persons interested in radio technique solely with a personal aim and without pecuniary interest.

§3.35 Amateur-Satellite Service: A radiocommunication service using space stations on earth satellites for the same purposes as those of the amateur service.

✶ ✶ ✶ ✶ ✶ ✶ ✶ ✶ ✶ ✶ ✶ ✶ ✶ ✶

Article 32—Amateur Service and Amateur-Satellite Service

Section I. Amateur Service

§1. Radiocommunications between amateur stations of different countries shall be forbidden if the administration of one of the countries concerned has notified that it objects to such radiocommunications.

§2. (1) When transmissions between amateur stations of different countries are permitted, they shall be made in plain language and shall be limited to messages of a technical nature relating to tests and to remarks of a personal character for which, by reason of their unimportance, recourse to the public telecommunications service is not justified.

(2) It is absolutely forbidden for amateur stations to be used for transmitting international communications on behalf of third parties.

(3) The preceding provisions may be modified by special arrangements between the administrations of the countries concerned.

§3. (1) Any person seeking a license to operate the apparatus of an amateur station shall prove that he is able to send correctly by hand and to receive correctly by ear, texts in Morse code signals. The administrations concerned may, however, waive this requirement in the case of stations making use exclusively of frequencies above 30 MHz.

(2) Administrations shall take such measures as they judge necessary to verify the operational and technical qualifications of any person wishing to operate the apparatus of an amateur station.

§4. The maximum power of amateur stations shall be fixed by the administrations concerned, having regard to the technical qualifications of the operators and to the conditions under which these stations are to operate.

§5. (1) All the general rules of the Convention and of these Regulations shall apply to amateur stations. In particular, the emitted frequency shall be as stable and as free from spurious emissions as the state of technical development for such stations permits.

(2) During the course of their transmissions, amateur stations shall transmit their call sign at short intervals.

Section II. Amateur-Satellite Service

§6. The provisions of Section I of this Article shall apply equally, as appropriate, to the Amateur-Satellite Service.

§7. Space stations in the Amateur-Satellite Service operating in bands shared with other services shall be fitted with appropriate devices for controlling emissions in the event that harmful interference is reported in accordance with the procedure laid down in Article 22. Administrations authorizing such space stations shall inform the IFRB and shall ensure that sufficient earth command stations are established before launch to guarantee that any harmful interference which might be reported can be terminated by the authorizing administration.

∗ ∗ ∗ ∗ ∗ ∗ ∗ ∗ ∗ ∗ ∗ ∗ ∗ ∗ ∗

Resolution No. 640

Relating to the International Use of Radiocommunications, in the Event of Natural Disasters, in Frequency Bands Allocated to the Amateur Service.

The World Administrative Radio Conference, Geneva, 1979,

Considering

a) that in the event of natural disaster normal communication systems are frequently overloaded, damaged, or completely disrupted;

b) that rapid establishment of communication is essential to facilitate worldwide relief actions;

c) that the amateur bands are not bound by international plans or notification procedures, and are therefore well adapted for short-term use in emergency cases;

d) that international disaster communications would be facilitated by temporary use of certain frequency bands allocated to the amateur service;

e) that under those circumstances the stations of the amateur service, because of their widespread distribution and their demonstrated capacity in such cases, can assist in meeting essential communication needs;

f) the existence of national and regional amateur emergency networks using frequencies throughout the bands allocated to the amateur service;

g) that in the event of a natural disaster, direct communication between amateur stations and other stations might enable vital communications to be carried out until normal communications are restored;

Recognizing

that the rights and responsibilities for communications in the event of a natural disaster rest with the administrations involved;

Resolves

1. that the bands allocated to the amateur service which are specified in No. 510 may be used by administrations to meet the needs of international disaster communications;

2. that such use of these bands shall be only for communications in relation to relief operations in connection with natural disasters;

3. that the use of specified bands allocated to the amateur service by non-amateur stations for disaster communications shall be limited to the duration of the emergency and to the specific geographical areas as defined by the responsible authority of the affected country;

4. that disaster communications shall take place within the disaster area and between the disaster area and the permanent headquarters of the organization providing relief;

5. that such communications shall be carried out only with the consent of the administration of the country in which the disaster has occurred;

6. that relief communications provided from outside the country in which the disaster has occurred shall not replace existing national or international amateur emergency networks;

7. that close cooperation is desirable between amateur stations and the stations of other radio services which may find it necessary to use amateur frequencies in disaster communications;

8. that such international relief communications shall avoid, as far as practicable, interference to the amateur service networks;

Invites Administrations

1. to provide for the needs of international disaster communications;

2. to provide for the needs of emergency communications with their national regulations.

* * * * * * * * * * * * * *

Resolution No. 641

Use of the Frequency Band 7000-7100 kHz

The World Administrative Radio Conference for the Planning of the HF Bands Allocated to the Broadcasting Service (Geneva, 1987),

Considering

a) that the sharing of frequency bands by amateur and broadcasting services is undesirable and should be avoided;

b) that it is desirable to have world-wide exclusive allocations for these services in Band 7;

c) that the band 7000-7100 kHz is allocated on a world-wide basis exclusively to the amateur service;

Resolves

that the broadcasting service shall be prohibited from the band 7000-7100 kHz and that the broadcasting stations operating on frequencies in this band shall cease such operation.

urges

the administrations responsible for the broadcasting stations operating on frequencies in the band 7000-7100 kHz to take the necessary steps to ensure that such operation ceases immediately, *instructs the Secretary-General*

to bring this Resolution to the attention of administrations.

✳ ✳ ✳ ✳ ✳ ✳ ✳ ✳ ✳ ✳ ✳ ✳ ✳ ✳ ✳

Resolution No. 642

Relating to the Bringing into Use of Earth Stations in the Amateur-Satellite Service.

The World Administrative Radio Conference, Geneva, 1979,

Recognizing

that the procedures of Articles 11 and 13 are applicable to the Amateur-Satellite Service;

Recognizing Further

a) that the characteristics of earth stations in the Amateur-Satellite Service vary widely;

b) that space stations in the Amateur-Satellite Service are intended for multiple access by amateur earth stations in all countries;

c) that coordination among stations in the amateur and Amateur-Satellite Services is accomplished without the need for formal procedures;

d) that the burden of terminating any harmful interference is placed upon the administration authorizing a space station in the Amateur-Satellite Service pursuant to the provisions of No. 2741 of the Radio Regulations;

Notes

that certain information specified in Appendices 3 and 4 cannot reasonably be provided for earth stations in the Amateur-Satellite Service;

Resolves

1. that when an administration (or one acting on behalf of a group of named administrations) intends to establish a satellite system in the Amateur-Satellite Service and wishes to publish information with respect to earth stations in that system it may:

1.1 communicate to the IFRB all or part of the information listed in Appendix 3; the IFRB shall publish such information in a special section of its weekly circular requesting comments to be communicated within a period of four months after the date of publication;

1.2 notify under Nos. 1488 to 1491 all or part of the information listed in Appendix 3; the IFRB shall record it in a special list;

2. that this information shall include at least the characteristics of a typical amateur earth station in the Amateur-Satellite Service having the facility to transmit signals to the space station to initiate, modify, or terminate the functions of the space station.

Appendix 6
PL 103-408—Joint Resolution of Congress to Recognize the Achievements of Radio Amateurs

PUBLIC LAW 103-408—OCT. 22, 1994

Public Law 103-408
103d Congress
Joint Resolution

To recognize the achievements of radio amateurs, and to establish support for such amateurs as national policy.

Whereas Congress has expressed its determination in section 1 of the Communications Act of 1934 (47 U.S.C. 151) to promote safety of life and property through the use of radio communication;

Whereas Congress, in section 7 of the Communications Act of 1934 (47 U.S.C. 157), established a policy to encourage the provision of new technologies and services;

Whereas Congress, in section 3 of the Communications Act of 1934, defined radio stations to include amateur stations operated by persons interested in radio technique without pecuniary interest;

Whereas the Federal Communications Commission has created an effective regulatory framework through which the amateur radio service has been able to achieve the goals of the service;

Whereas these regulations, set forth in Part 97 of title 47 of the Code of Federal Regulations clarify and extend the purposes of the amateur radio service as a—

(1) voluntary noncommercial communication service, particularly with respect to providing emergency communications;

(2) contributing service to the advancement of the telecommunications infrastructure;

(3) service which encourages improvement of an individual's technical and operating skills;

(4) service providing a national reservoir of trained operators, technicians and electronics experts; and

(5) service enhancing international good will;

Whereas Congress finds that members of the amateur radio service community has provided invaluable emergency communications services following such disasters as Hurricanes Hugo, Andrew, and Iniki, the Mt. St. Helens Eruption, the Loma Prieta earthquake, tornadoes, floods, wild fires, and industrial accidents in great number and variety across the Nation; and

Whereas Congress finds that the amateur radio service has made a contribution to our Nation's communications by its crafting, in 1961, of the first Earth satellite licensed by the Federal Communications Commission, by its proof-of-concept for search rescue satellites, by its continued exploration of the low Earth orbit in particular pointing the way to commercial use thereof in the 1990s, by its pioneering of communications using reflections from meteor trails, a technique now used for certain government and commercial communications, and by its leading role in development of low-cost, practical data transmission by radio which increasingly is being put to extensive use in, for instance, the land mobile service: Now, therefore, be it

Resolved by the Senate and House of Representatives of the United States of America in Congress assembled,

SECTION 1. FINDINGS AND DECLARATIONS OF CONGRESS

Congress finds and declares that—

(1) radio amateurs are hereby commended for their contributions to technical progress in electronics, and for their emergency radio communications in times of disaster;

(2) the Federal Communications Commission is urged to continue and enhance the development of the amateur radio service as a public benefit by adopting rules and regulations which encourage the use of new technologies within the amateur radio service; and

(3) reasonable accommodation should be made for the effective operation of amateur radio from residences, private vehicles and public areas, and that regulation at all levels of government should facilitate and encourage amateur radio operation as a public benefit.

Approved October 22, 1994.

Appendix 7
Memorandum Opinion and Order in PRB-1

Here's the full text of FCC's Memorandum Opinion and Order in PRB-1. If you require an "official" copy of PRB-1 for use in a legal proceeding, you may cite the *Federal Register:* 50 FR 38813.

Before the
Federal Communications Commission   FCC 85-506
Washington, DC 20554   36149

In the Matter of   )
)
Federal preemption of state and   )   PRB-1
local regulations pertaining   )
to Amateur radio facilities.   )

MEMORANDUM OPINION AND ORDER

Adopted: September 16, 1985   ;   Released: September 19, 1985

By the Commission: Commissioner Rivera not participating.

Background

1. On July 16, 1984, the American Radio Relay League, Inc (ARRL) filed a Request for Issuance of a Declaratory Ruling asking us to delineate the limitations of local zoning and other local and state regulatory authority over Federally-licensed radio facilities. Specifically, the ARRL wanted an explicit statement that would preempt all local ordinances which provably preclude or significantly inhibit effective reliable amateur radio communications. The ARRL acknowledges that local authorities can regulate amateur installations to insure the safety and health of persons in the community, but believes that those regulations cannot be so restrictive that they preclude effective amateur communications.

2. Interested parties were advised that they could file comments in the matter.[1] With extension, comments were due on or before December 26, 1984,[2] with reply comments due on or before January 25, 1985.[3] Over sixteen hundred comments were filed.

Local Ordinances

3. Conflicts between amateur operators regarding radio antennas and local authorities regarding restrictive ordinances are common. The amateur operator is governed by the regulations contained in Part 97 of our rules. Those rules do not limit the height of an amateur antenna but they require, for aviation safety reasons, that certain FAA notification and FCC approval procedures must be followed for antennas which exceed 200 feet in height above ground level or antennas which are to be erected near airports. Thus, under FCC rules some antenna support structures require obstruction marking and lighting. On the other hand, local municipalities or governing bodies frequently enact regulations limiting antennas and their support structures in height and location, e.g. to side or rear yards, for health, safety or aesthetic considerations. These limiting regulations can result in conflict because the effectiveness of the communications that emanate from an amateur radio station are directly dependent upon the location and the height of the antenna. Amateur operators maintain that they are precluded from operating in certain bands allocated for their use if the height of their antennas is limited by a local ordinance.

4. Examples of restrictive local ordinances were submitted by several amateur operators in this proceeding. Stanley J. Cichy, San Diego, California, noted that in San Diego amateur radio antennas come under a structures ruling which limits building heights to 30 feet. Thus, antennas there are also limited to 30 feet. Alexander Vrenios, Mundelein, Illinois, wrote that an ordinance of the Village of Mundelein provides that an antenna must be a distance from the property line that is equal to one and one-half times its height. In his case, he is limited to an antenna tower for his amateur station just over 53 feet in height.

5. John C. Chapman, an amateur living in Bloomington, Minnesota, commented that he was not able to obtain a building permit to install an amateur radio antenna exceeding 35 feet in height because the Bloomington city ordinance restricted "structures" heights to 35 feet. Mr. Chapman said that the ordinance, when written, undoubtedly applied to buildings but was now being applied to antennas in the absence of a specific ordinance regulating them. There were two options open to him if he wanted to engage in amateur communications. He could request a variance to the ordinance by way of a hearing before the City Council, or he could obtain affidavits from his neighbors swearing that they had no objection to the proposed antenna installation. He got the building permit after obtaining the cooperation of his neighbors. His concern, however, is that he had to get permission from several people before he could effectively engage in radio communications for which he had a valid FCC amateur license.

6. In addition to height restrictions, other limits are enacted by local jurisdictions—anti-climb devices on towers or fences around them; minimum distances from high voltage power lines; minimum distances of towers from property lines; and regulations pertaining to the structural soundness of the antenna installation. By and large, amateurs do not find these safety precautions objectionable. What they do object to are the sometimes prohibitive, non-refundable application filing fees to obtain a permit to erect an antenna installation and those provisions in ordinances which regulate antennas for purely aesthetic reasons. The amateurs contend, almost universally, that "beauty is in the eye of the beholder." They assert that an antenna installation is not more aesthetically displeasing than other objects that people keep on their property, e.g. motor homes, trailers, pick-up trucks, solar collectors and gardening equipment.

Restrictive Covenants

7. Amateur operators also oppose restrictions on their amateur operations which are contained in the deeds for their homes or in their apartment leases. Since these restrictive covenants are contractual agreements between private parties, they are not generally a matter of concern to the Commission. However, since some amateurs who commented in this proceeding provided us with examples of restrictive covenants, they are included for information. Mr. Eugene O. Thomas of Hollister, California, included in his comments an extract of the Declaration of Covenants and Restrictions for Ridgemark Estates, County of San Benito, State of California. It provides:

> No antenna for transmission or reception of radio signals shall be erected outdoors for use by any dwelling unit except upon approval of the Directors. No radio or television signals or any other form of electromagnetic radiation shall be permitted to originate from any lot which may unreasonably interfere with the reception of television or radio signals upon any other lot.

Marshall Wilson, Jr. provided a copy of the restrictive covenant contained in deeds for the Bell Martin Addition #2, Irving, Texas. It is binding upon all of the owners or purchasers of the lots in the said addition, his or their heirs, executors, administrators or assigns. It reads:

> No antenna or tower shall be erected upon any lot for the purposes of radio operations.

William J. Hamilton resides in an apartment building in Gladstone, Missouri. He cites a clause in his lease prohibiting the erection of an antenna. He states that he has been forced to give up operating amateur radio equipment except a hand-held 2 meter (144-148 MHz) radio transceiver. He maintains that he should not be penalized just because he lives in an apartment.

Other restrictive covenants are less global in scope than those cited above. For example, Robert Webb purchased a home in Houston, Texas. His deed restriction prohibited "transmitting or receiving antennas extending above the roof line."

8. Amateur operators generally oppose restrictive covenants for several reasons. They maintain that such restrictions limit the places that they can reside if they want to pursue their hobby of amateur radio. Some state that they impinge on First Amendment rights of speech. Others believe that a constitutional right is being abridged because, in their view, everyone has a right to access the airwaves regardless of where they live.

9. The contrary belief held by housing subdivision communities and condominium or homeowner's associations is that amateur radio installations constitute safety hazards, cause interference to other electronic equipment which may be operated in the home (television, radio, stereos) or are eyesores that detract from the aesthetic and tasteful appearance of the housing development or apartment complex. To counteract these negative consequences, the subdivisions and associations include in their deeds, leases or by-laws, restrictions and limitations on the location and height of antennas or, in some cases, prohibit them altogether. The restrictive covenants are contained in the contractual agreement entered into at the time of the sale or lease of the property. Purchasers or lessees are free to choose whether they wish to reside where such restrictions on amateur antennas are in effect or settle elsewhere.

Supporting Comments

10. The Department of Defense (DOD) supported the ARRL and emphasized

in its comments that continued success of existing national security and emergency preparedness telecommunications plans involving amateur stations would be severely diminished if state and local ordinances were allowed to prohibit the construction and usage of effective amateur transmission facilities. DOD utilizes volunteers in the Military Affiliate Radio Service (MARS),[4] Civil Air Patrol (CAP) and the Radio Amateur Civil Emergency Service (RACES). It points out that these volunteer communicators are operating radio equipment installed in their homes and that undue restrictions on antennas by local authorities adversely affect their efforts. DOD states that the responsiveness of these volunteer systems would be impaired if local ordinances interfere with the effectiveness of these important national telecommunication resources. DOD favors the issuance of a ruling that would set limits for local and state regulatory bodies when they are dealing with amateur stations.

11. Various chapters of the American Red Cross also came forward to support the ARRL's request for a preemptive ruling. The Red Cross works closely with amateur radio volunteers. It believes that without amateurs' dedicated support, disaster relief operations would significantly suffer and that its ability to serve disaster victims would be hampered. It feels that antenna height limitations that might be imposed by local bodies will negatively affect the service now rendered by the volunteers.

12. Cities and counties from various parts of the United States filed comments in support of the ARRL's request for a Federal preemption ruling. The comments from the Director of Civil Defense, Port Arthur, Texas, are representative:

> The Amateur Radio Service plays a vital role with our Civil Defense program here in Port Arthur and the design of these antennas and towers lends greatly to our ability to communicate during times of disaster.
>
> We do not believe there should be any restrictions on the antennas and towers except for reasonable safety precautions. Tropical storms, hurricanes and tornadoes are a way of life here on the Texas Gulf Coast and good communications are absolutely essential when preparing for a hurricane and even more so during recovery operations after the hurricane has past.

13. The Quarter Century Wireless Association took a strong stand in favor of the Issuance of a declaratory ruling. It believes that Federal preemption is necessary so that there will be uniformity for all Amateur Radio installations on private property throughout the United States.

14. In its comments, the ARRL argued that the Commission has the jurisdiction to preempt certain local land use regulations which frustrate or prohibit amateur radio communications. It said that the appropriate standard in preemption cases is not the extent of state and local interest in a given regulation, but rather the impact of the regulation on Federal goals. Its position is that Federal preemption is warranted whenever local government regulations relate adversely to the operational aspects of amateur communication. The ARRL maintains that localities routinely employ a variety of land use devices to preclude the installation of effective amateur antennas, including height restrictions, conditional use permits, building setbacks and dimensional limitations on antennas. It sees a declaratory ruling of Federal preemption as necessary to cause municipalities to accommodate amateur operator needs in land use planning efforts.

15. James C. O'Connell, an attorney who has represented several amateurs before local zoning authorities, said that requiring amateurs to seek variances or

special use approval to erect reasonable antennas unduly restricts the operation of amateur stations. He suggested that the Commission preempt zoning ordinances which impose antenna height limits of less than 65 feet. He said that this height would represent a reasonable accommodation of the communication needs of most amateurs and the legitimate concerns of local zoning authorities.

Opposing Comments

16. The City of La Mesa, California, has a zoning regulation which controls amateur antennas. Its comments reflected an attempt to reach a balanced view.

> This regulation has neither the intent, nor the effect, of precluding or inhibiting effective and reliable communications. Such antennas may be built as long as their construction does not unreasonably block views or constitute eyesores. The reasonable assumption is that there are always alternatives at a given site for different placement, and/or methods for aesthetic treatment. Thus, both public objectives of controlling land use for the public health, safety, and convenience, and providing an effective communications network, can be satisfied. A blanket to completely set aside local control, or a ruling which recognizes control only for the purpose of safety of antenna construction, would be contrary to…legitimate local control.

17. Comments from the County of San Diego state:

> While we are aware of the benefits provided by amateur operators, we oppose the issuance of a preemption ruling which would elevate 'antenna effectiveness' to a position above all other considerations. We must, however, argue that the local government must have the ability to place reasonable limitations upon the placement and configuration of amateur radio transmitting and receiving antennas. Such ability is necessary to assure that the local decision-makers have the authority to protect the public health, safety and welfare of all citizens.
>
> In conclusion, I would like to emphasize an important difference between your regulatory powers and that of local governments. Your Commission's approval of the preemptive requests would establish a "national policy." However, any regulation adopted by a local jurisdiction could be overturned by your Commission or a court if such regulation was determined to be unreasonable.

18. The City of Anderson, Indiana, summarized some of the problems that face local communities:

> I am sympathetic to the concerns of these antenna owners and I understand that to gain the maximum reception from their devices, optimal location is necessary. However, the preservation of residential zoning districts as "liveable" neighborhoods is jeopardized by placing these antennas in front yards of homes. Major problems of public safety have been encountered, particularly vision blockage for auto and pedestrian access. In addition, all communities are faced with various building lot sizes. Many building lots are so small that established setback requirements (in order to preserve adequate air and light) are vulnerable to the unregulated placement of antennas. …the exercise of preemptive authority by the FCC in granting this request would not be in the best interest of the general public.

19. The National Association of Counties (NACO), the American Planning Association (APA) and the National League of Cities (NLC) all opposed the issuance of an antenna preemption ruling. NACO emphasized that federal and state power must be viewed in harmony and warns that Federal intrusion into local concerns of health, safety and welfare could weaken the traditional police power

exercised by the state and unduly interfere with the legitimate activities of the states. NLC believed that both Federal and local interests can be accommodated without preempting local authority to regulate the installation of amateur radio antennas. The APA said that the FCC should continue to leave the issue of regulating amateur antennas with the local government and with the state and Federal courts.

Discussion

20. When considering preemption, we must begin with two constitutional provisions. The tenth amendment provides that any powers which the constitution either does not delegate to the United States or does not prohibit the states from exercising are reserved to the states. These are the police powers of the states. The Supremacy Clause, however, provides that the constitution and the laws of the United States shall supersede any state law to the contrary. Article III, Section 2. Given these basic premises, state laws may be preempted in three ways: First, Congress may expressly preempt the state law. See *Jones v. Rath Packing Co.*, 430 U.S. 519, 525 (1977). Or, Congress may indicate its intent to completely occupy a given field so that any state law encompassed within that field would implicitly be preempted. Such intent to preempt could be found in a congressional regulatory scheme that was so pervasive that it would be reasonable to assume that Congress did not intend to permit the states to supplement it. See *Fidelity Federal Savings & Loan Ass'n v. de la Cuesta,* 458 U.S. 141, 153 (1982). Finally, preemption may be warranted when state law conflicts with federal law. Such conflicts may occur when "compliance with both Federal and state regulations is a physical impossibility," *Florida Lime & Avocado Growers, Inc. v. Paul,* 373 U.S. 132, 142, 143 (1963), or when state law "stands as an obstacle to the accomplishment and execution of the full purposes and objectives of Congress," *Hines v. Davidowitz,* 312 U.S. 52, 67 (1941). Furthermore, federal regulations have the same preemptive effect as federal statues, *Fidelity Federal Savings & Loan Association v. de la Cuesta,* supra.

21. The situation before us requires us to determine the extent to which state and local zoning regulations may conflict with federal policies concerning amateur radio operators.

22. Few matters coming before us present such a clear dichotomy of view point as does the instant issue. The cities, countries, local communities and housing associations see an obligation to all of their citizens and try to address their concerns. This is accomplished through regulations, ordinances or covenants oriented toward the health, safety and general welfare of those they regulate. At the opposite pole are the individual amateur operators and their support groups who are troubled by local regulations which may inhibit the use of amateur stations or, in some instances, totally preclude amateur communications. Aligned with the operators are such entities as the Department of Defense, the American Red Cross and local civil defense and emergency organizations who have found in Amateur Radio a pool of skilled radio operators and a readily available backup network. In this situation, we believe it is appropriate to strike a balance between the federal interest in promoting amateur operations and the legitimate interests of local governments in regulating local zoning matters. The cornerstone on which we will predicate our decision is that a reasonable accommodation may be made between the two sides.

23. Preemption is primarily a function of the extent of the conflict between federal and state and local regulation. Thus, in considering whether our regulations or policies can tolerate a state regulation, we may consider such factors as the severity of the conflict and the reasons underlying the state's regulations. In this regard, we have previously recognized the legitimate and important state interests reflected in local zoning regulations. For example, in *Earth Satellite Communications, Inc.*, 95 FCC 2d 1223 (1983), we recognized that

> ...countervailing state interests inhere in the present situation...For example, we do not wish to preclude a state or locality from exercising jurisdiction over certain elements of an SMATV operation that properly may fall within its authority, such as zoning or public safety and health, provided the regulation in question is not undertaken as a pretext for the actual purpose of frustrating achievement of the preeminent federal objective and so long as the non-federal regulation is applied in a nondiscriminatory manner.

24. Similarly, we recognize here that there are certain general state and local interests which may, in their even-handed application, legitimately affect amateur radio facilities. Nonetheless, there is also a strong federal interest in promoting amateur communications. Evidence of this interest may be found in the comprehensive set of rules that the Commission has adopted to regulate the amateur service. [5] Those rules set forth procedures for the licensing of stations and operators, frequency allocations, technical standards which amateur radio equipment must meet and operating practices which amateur operators must follow. We recognize the amateur radio service as a voluntary, noncommercial communication service, particularly with respect to providing emergency communications. Moreover, the amateur radio service provides a reservoir of trained operators, technicians and electronic experts who can be called on in times of national or local emergencies. By its nature, the Amateur Radio Service also provides the opportunity for individual operators to further international goodwill. Upon weighing these interests, we believe a limited preemption policy is warranted. State and local regulations that operate to preclude amateur communications in their communities are in direct conflict with federal objectives and must be preempted.

25. Because amateur station communications are only as effective as the antennas employed, antenna height restrictions directly affect the effectiveness of amateur communications. Some amateur antenna configurations require more substantial installations than others if they are to provide the amateur operator with the communications that he/she desires to engage in. For example, an antenna array for international amateur communications will differ from an antenna used to contact other amateur operators at shorter distances. We will not, however, specify any particular height limitation below which a local government may not regulate, nor will we suggest the precise language that must be contained in local ordinances, such as mechanisms for special exceptions, variances, or conditional use permits. Nevertheless, local regulations which involve placement, screening, or height of antennas based on health, safety, or aesthetic considerations must be crafted to accommodate reasonably amateur communications, and to represent the minimum practicable regulation to accomplish the local authority's legitimate purpose.[6]

26. Obviously, we do not have the staff or financial resources to review all state and local laws that affect amateur operations. We are confident, however, that state and local governments will endeavor to legislate in a manner that affords appropriate recognition to the important federal interest at stake here and thereby avoid

unnecessary conflicts with federal policy, as well as time-consuming and expensive litigation in this area. Amateur operators who believe that local or state governments have been overreaching and thereby have precluded accomplishment of their legitimate communications goals, may, in addition, use this document to bring our policies to the attention of local tribunals and forums.

27. Accordingly, the Request for Declaratory Ruling filed July 16, 1984, by the American Radio Relay League, Inc., IS GRANTED to the extent indicated herein and in all other respects, IS DENIED.

FEDERAL COMMUNICATIONS COMMISSION
William J. Tricarico
Secretary

Footnotes

[1]Public Notice, August 30, 1984, Mimeo. No. 6299, 49 F.R. 36113, September 14, 1984.

[2]Public Notice, December 19,1984, Mimeo. No. 1498.

[3]Order, November 8, 1984, Mimeo, No. 770.

[4]MARS is solely under the auspices of the military which recruits volunteer amateur operators to render assistance to it. The Comission is not involved in the MARS program.

[5]47 CFR Part 97.

[6]We reiterate that our ruling herein does not reach restrictive covenants in private contractual agreements. Such agreements are voluntarily entered into by the buyer or tenant when the agreement is executed and do not usually concern this Commission.

Appendix 8
FCC Letter: Local Authorities May Not Regulate RFI

FEDERAL COMMUNICATIONS COMMISSION
WASHINGTON, D.C. 20554
FEB 14, 1990

Christopher D. Imlay, Esquire
American Radio Relay League, Inc.
Office of Legal Counsel
Washington, D.C.

Re: Ordinance Regulating Radio Frequency Interference, Pierre, South Dakota

Dear Mr. Imlay:

This is in response to your letter of January 16, 1990, concerning an ordinance enacted in Pierre, South Dakota, empowering the City Inspector to investigate and prohibit emissions by radios and other electronic devices which cause or create interference to television or radio reception. You state that the City Inspector has enforced this ordinance against an amateur radio operator licensed by the Commission, and you seek an opinion concerning the validity of the ordinance.

Congress has preempted any concurrent state or local regulation of radio interference pursuant to the provisions of the Communications Act. See 47 U.S.C. §302(a). Section 302(a)(1) of the Act provides that the "Commission may, consistent with the public interest, convenience, and necessity, make reasonable regulations (1) governing the interference potential of devices which in their operation are capable of emitting radio frequency energy by radiation, conduction, or other means in sufficient degree to cause harmful interference to radio communications . . . ." 47 U.S.C. §302(a)(1). The legislative history of Section 302(a) provides explicitly that the Commission has exclusive authority to regulate radio frequency interference (RFI). In its Conference Report No. 97-765, Congress declared:

> The Conference Substitute is further intended to clarify the reservation of exclusive jurisdiction to the Federal Communications Commission over matters involving RFI. Such matters shall not be regulated by local or state law, nor shall radio

transmitting be subject to local or state regulation as part of any effort to resolve an RFI complaint.

H.R. Report No. 765, 97th Cong., 2d Sess. 33 (1982), reprinted at 1982 U.S. Code Cong. & Ad News 2277.

State laws that require amateurs to cease operations or incur penalties as a consequence of radio interference thus have been entirely preempted by Congress.

Of course, any member of the public may seek the Commission's assistance in resolving interference problems. The Commission's Field Operations Bureau (FOB) frequently investigates radio interference complaints and has prepared the enclosed pamphlets describing the various remedies available to address radio interference matters. Members of the public in Pierre experiencing interference may also wish to contact Dennis P. Carlton, Engineer-in-Charge of FOB's Denver Office at (303) 236-8026.

I trust the foregoing is responsive to your inquiry.

Sincerely yours,

Robert L. Pettit
General Counsel

Enclosures

cc: City Inspector, Pierre, South Dakota

Appendix 9
FCC Letter: Local Authorities May Not Deny Tower Based on RFI

FEDERAL COMMUNICATIONS COMMISSION
WASHINGTON, D.C. 20554
25 OCT 1994

IN REPLY REFER TO:
7240-F/1700C1

Board of Zoning Appeals
Town of Hempstead
1 Washington Street
Hempstead, New York, 11550-4923

Dear Board Members:

It has come to our attention that the Town of Hempstead's Board of Zoning Appeals (Board) has denied Mr. Hayden M. Nadel's application for a variance permitting him to maintain his amateur radio station's antenna at a height of fifty-five feet (versus the thirty feet permitted by the zoning ordinance). According the text of the Board's decision (provided by Mr. Nadel), it based its determination largely on its finding that the "proposed and existing antenna height of fifty-five feet" was resulting in interference to the home electronic equipment of Mr. Nadel's neighbors.

Local Governments must reasonably accommodate amateur operations in zoning decisions. See PRB-1, 101 FCC 2d 952 (1985) and Section 97.15(e) of the Commission's Rules, 47 C.F.R. §97.15(e) (copies enclosed). Section 97.15(e) provides that an amateur station antenna structure may be erected at heights and dimensions sufficient to accommodate amateur service communications. Local authorities may adopt regulations pertaining to placement, screening, or height of antennas, if such regulations are based on health, safety, or aesthetic considerations and reasonably accommodate amateur communications. They may not, however, base their regulation of amateur service antenna structures on the causation of interference to home electronic equipment —an area regulated exclusively by the Commission.

The Commission's jurisdiction over interference matters is set forth in Section 302(a) of the Communications Act of 1934, as amended, 47 U.S.C. §302(a) (copy enclosed). It is clear from the report of the Joint Committee of Conference, H.R. Report No. 765, 97th Cong., 2nd Sess. (pertinent excerpts enclosed), that the congress intended that the Commission have exclusive jurisdiction over interference to home electronic equipment.

I would also like to point out that there is no reasonable connection between requiring Mr. Nadel to reduce the height of his antenna and reducing the amount of interference to his neighbor's home electronic equipment. On the contrary, antenna height is inversely related to the strength, in the horizontal plane, of the radio signal that serves as a catalyst for interference in susceptible home electronic equipment. It is a matter of technical fact that the higher an amateur antenna, the less likely it is that radio frequency interference will appear in home electronic equipment.

I hope the information in this letter is helpful

        Sincerely,

        Ralph A. Haller
        Chief, Private Radio Bureau

Enclosures

Appendix 10
Chronological List of Landmark Amateur Radio Case Cites (Antennas & RFI cases)

Schroeder v. Municipal Court of Cerritos
   73 Cal. Rptr 3d 841, 141 Cal. Rptr. 85 (1977)
   Appeal dismissed 435 US 990 (1978)

Guske v. Oklahoma City, OK
   763 F. 2d 379 (10th Cir 1985)

PRB-1 Declaratory Ruling
   Amateur Radio Preemption, 101 FCC 2d 952 (1985)

Satellite Receive Only Earth Stations Preemption Order
   Docket 85-87; Pike and Fischer Radio Regulation 2d 1073 (1986)

John Thernes v. City of Lakeside Park, Kentucky, et al
   779 F. 2d 1187 (6th Cir. 1986)
   Final Judgment; 62 Pike and Fischer Radio Regulation 2d, 284 (E. D. KY, 1987)

Andrew B. Bodony v. Incorporated Village of Sands Point et al (NY)
   681 F. Supp 1009 (E.D. NY 1987)

William F. Bulchis v. City of Edmonds (WA)
   671 F. Supp 1270 (W. D. Wash 1987)

Izzo v. Borough of River Edge, New York, et al
   843 F. 2d 765 (3rd Cir., 1988)

James D. MacMillan v. City of Rocky River, Ohio, et al
   748 F. Supp 1241 (N.D.Ohio 1990)

John F. Williams v. City of Columbia, SC
   707 F. Supp 207 (D.S.C., 1989)
   Affirmed 906 F.2d 994 (4th Cir. 1990)

Vernon Howard v. City of Burlingame (CA)
   726 F. Supp 770 (N.D.Cal. 1989)
   Affirmed 937 F. 2d 1376 9th Cir. 1991

Palmer Hotz et al v. James E. Rich (Cov.)
   6 Cal. Rptr. 2d 219 (Cal. Ct. of Appeals 1992)

D. R. Evans v. Board of County Commissioners of the County of Boulder,
   Colorado, et al
   752 F. Supp 973 (D.Colo. 1990)
   Reversed 994 F. 2d 761 (10th Cir 1993)

Sylvia Pentel v. City of Mendota Heights (MN)
   13 F. 3d 1261 (8th Cir 1994)

Appendix 11
Docket 91-36, Scanner Preemption

Before the Federal
Communications Commission
Washington, D.C. 20554

PR Docket No. 91-36

In the Matter of

Federal Preemption of State
and Local Laws Concerning Amateur
Operator Use of Transceivers
Capable of Reception Beyond
Amateur Service Frequency
Allocations

MEMORANDUM OPINION AND ORDER

Adopted: August 20, 1993;

Released: September 3, 1993

By the Commission:

I. INTRODUCTION

1. On November 14, 1989, the American Radio Relay League, Incorporated (ARRL), filed a *Motion for a Declaratory Ruling*[1] requesting that the Commission preempt certain state statutes and local ordinances affecting transceivers[2] used by Amateur Radio Service Licensees. The laws referenced by the ARRL prohibit the possession of such transceivers if they are capable of the reception of communications on certain frequencies other than amateur service frequencies. On March 15, 1990, we released a public notice[3] inviting comment on ARRL's request. In addition, on February 28, 1991, we released a *Notice of Inquiry*[4] that solicited additional comment to assist us in making a decision in this matter. This *Memorandum Opinion and Order* grants the request to the extent indicated herein.

II. BACKGROUND

2. The ARRL motion discusses state statutes and local ordinances commonly

known as "scanner laws," the violation of which may be a criminal misdemeanor with the possibility of equipment confiscation.[5] Specifically, ARRL notes that state statutes in New Jersey and Kentucky (which have subsequently been changed —see paragraph 3, *infra*) prohibit the possession of a mobile short-wave radio capable of receiving frequencies assigned by the Commission for, *inter alia*, police use.[6] In addition, ARRL states that local ordinances exist throughout the United States that similarly prohibit the possession of such mobile short-wave radios without a locally-issued permit.[7] Therefore, ARRL explains, scanner laws can, *inter alia*, render amateur radio licensees traveling interstate by automobile vulnerable to arrest and to the seizure of their radio equipment by state or local police.[8]

3. Since the ARRL motion was filed with the Commission, New Jersey repealed its statute and substituted a new, narrowly tailored scanner law that only applies in the criminal context.[9] In addition, Kentucky amended its statute by adding an exemption applying to amateur radio licensees.[10] As a result, there no longer appears to be any state scanner law with a deleterious effect on the legitimate operations of amateur radio service licensees. Nonetheless, the preemption issue raised by the ARRL motion remains timely because it appears that some local scanning ordinances remain in effect without safeguards to protect the legitimate use of such radios by our licensees.[11]

III. MOTION, INQUIRY AND COMMENTS

A. The ARRL Motion.

4. ARRL makes two arguments in support of preemption. First, it states that the receiver sections of the majority of commercially available amateur station transceivers can be tuned slightly past the edges of the amateur service bands to facilitate adequate reception up to the end of the amateur service bands. ARRL seeks a preemption ruling that would permit amateur operators to install in vehicles transceivers that are capable of this "incidental" reception.[12] Although ARRL's formal request is couched in terms of this first, technical point, the request focuses almost entirely on a second, broader issue of whether state and local authorities should be permitted, via the scanner laws, to prohibit the capability of radio reception by amateur operators on public safety and special emergency frequencies that are well outside the amateur service bands.

5. Concerning the broader issue, ARRL argues that amateur operators have special needs for broadscale "out-of-band" reception, and that the marketplace has long recognized these needs by offering accommodating transceivers. According to ARRL,[13] all commercially manufactured amateur service HF transceivers and the majority of such VHF and UHF transceivers have non-amateur service frequency reception capability well beyond the "incidental" — they can receive across a broad spectrum of frequencies, including the police and other public safety and special emergency frequencies here at issue. This additional capability, argues ARRL, permits amateur operators to participate in a variety of safety activities, some in conjunction with the military or the National Weather Service. In both cases, reception on non-amateur frequencies is necessary. Such activities benefit the public, according to ARRL, especially in times of emergency,[14] and some require the mobile use of the amateur stations.[15] ARRL states that, in addition, the vast majority of amateur operators take part in these mobile activities, and that the widespread enforcement of scanner laws would render illegal the possession of

virtually all modern amateur mobile equipment.[16] ARRL states that, as a result of scanner laws, "several dozen instances of radio seizure and criminal arrest [have been] suffered by licensed amateurs."[17]

B. The Inquiry and Comments.

6. The Commission's February 28, 1991 *Inquiry* solicited additional information concerning the technical and financial feasibility of modifying existing amateur service mobile transceivers to render them incapable of receiving police or other public safety channels. We also asked for information concerning the current and future marketplace availability of mobile equipment meeting the restrictions of the laws and whether there is value in having an available pool of wide-band, mobile amateur equipment in the United States to meet emergency needs.

7. In response to the *Inquiry,* we received 115 comments and reply comments, of which the great majority are from individual amateurs who support the preemption.[18] One commenter, the Michigan Department of State Police, states that although it cooperates with the amateur service during emergencies, it is concerned about isolated incidents of apparently unlawful actions taken by amateur licensees upon receipt of public safety communications outside of the amateur radio band.[19] Therefore, it concludes that "there can be no beneficial need for amateur radio equipment to tune in public safety channels."[20] Of the remaining comments received, only a few address the technical and marketplace questions described above. These comments are from individual amateur operators[21] who state that existing wide-band transceivers cannot be modified to meet the restrictions of the scanner laws without substantial expense and that this situation will continue as new equipment becomes available. Despite our specific request in the *Inquiry* that manufacturers comment on these technical and financial questions, no manufacturer chose to respond on these points. We also received a few comments describing the prevalence of scanner laws nationwide.[22] Finally, the National Communications System (NCS), of the Department of Defense, states in its comment that the federal government utilizes amateur operators in a number of programs requiring mobile, wide-band transceivers.[23]

IV. DISCUSSION

8. There are three ways state and local laws may be preempted. First, Congress may expressly preempt the state or local law. Second, Congress may, through legislation, clearly indicate its intent to occupy the field of regulation, leaving "no room for the States to supplement."[24] Last, and most important for this discussion,

> [e]ven where Congress has not completely displaced state regulation in a specific area, state law [may be] nullified to the extent that it actually conflicts with federal law. Such a conflict arises when "compliance with both federal and state regulations is a physical impossibility,"...or when state law "stands as an obstacle to the accomplishment and execution of the full purposes and objectives of Congress." [25]

Furthermore, "[f]ederal regulations have no less preemptive effect than federal statutes."[26]

9. The amateur service is regulated extensively under Part 97 of the Commission's Rules, 47 C.F.R. Part 97. As we have stated in the past:

> [T]here is...a strong federal interest in promoting amateur communications. Evidence of this interest may be found in the comprehensive set of rules that the Commission has adopted to regulate the amateur service. Those rules set forth

procedures for the licensing of stations and operators, frequency allocations, technical standards which amateur radio equipment must meet and operating practices which amateur operators must follow. We recognize the Amateur Radio Service as a voluntary, noncommercial communication service, particularly with respect to providing emergency communications. Moreover, the Amateur Radio Service provides a reservoir of trained operators, technicians and electronic experts who can be called on in times of national or local emergencies. By its nature, the Amateur Radio Service also provides the opportunity for individual operators to further international goodwill.[27]

This federal interest in the amateur service is also reflected in Section 97.1 of our rules, 47 C.F.R. §97.1, which provides that the amateur service exists to "continu[e] and exten[d]...the amateur's proven ability to contribute to the advancement of the radio art."[28] This regulatory purpose is consistent with the Communications Act requirement that "[i]t shall be the policy of the United States to encourage the provision of new technologies to the public."[29]

10. The strong federal interest in the preservation and advancement of the amateur service is also demonstrated by Congress's recent recognition of the goals of the amateur service in a "Sense of Congress" provision in which Congress strongly encouraged and supported the amateur service.[30] Congress therein directed all Government agencies to take into account the valuable contribution of amateurs when considering actions affecting the amateur radio service.[31] We believe that the strong federal interest in supporting the emergency services provided by amateurs cannot be fully accomplished unless amateur operators are free to own and operate their stations to the fullest extent permitted by their licenses and are not unreasonably hampered in their ability to transport their radio transmitting stations across state and local boundaries for purposes of transmitting and receiving on authorized frequencies. Indeed, as a result of advances in technology making smaller, lighter weight radios commercially available, the Commission has expressly amended its rules to facilitate and encourage unrestricted mobile amateur operations. As we noted in a recent rule making proceeding to modify the rules governing the amateur radio service,

> In the age of the microprocessor and the integrated circuit [amateur] equipment is highly portable. It is common for amateur operators to carry hand-held transceivers capable of accessing many local repeaters in urban areas and also capable of reasonably good line-of-sight communication. It appears that the concept of fixed station operation no longer carries with it the same connotation it did previously. For this reason, we propose to delete current rules that relate to station operation away from the authorized fixed station location.[32]

As a consequence of these changes, the rules now expressly authorize amateur service operation "at points where the amateur service is regulated by the FCC," that is, at fixed and mobile locations throughout the United States. Furthermore, the Commission's Rules do not in any way prohibit an amateur service transceiver from having out-of-band reception capability.[33]

11. Against this background, we conclude that certain state and local laws, as described below, conflict with the Commission's regulatory scheme designed to promote a strong amateur radio service. Scanner laws that prohibit the use of transceivers that transmit and receive amateur frequencies because they also receive public safety, special emergency or other radio service frequencies frustrate most legitimate amateur service mobile operations through the threat of penalties

such as fines and the confiscation of equipment. As noted by ARRL, virtually all modern amateur service equipment in use today can receive transmissions on the public safety and special emergency frequencies at issue, and the majority of amateur stations[34] are operated in a mobile fashion. Consequently, the mobile operations of the vast majority of amateurs are affected by such laws. In addition, the record statements by amateurs that the costs would be substantial to modify existing transceivers are unchallenged. The scanner laws, then, essentially place the amateur operator in the position of either foregoing mobile operations by simply avoiding all use of the equipment in vehicles or other locations specified in the laws, or risking fines, or equipment confiscation. This very significant limitation on amateurs operating rights runs counter to the express policies of both Congress and the Commission to encourage and support amateurs service operations, including mobile operations, and impermissibly encroaches on federal authority over amateur operators.[35] It conflicts directly with the federal interest in amateur operators being able to transmit and receive on authorized amateur service frequencies.[36]

12. For these reasons, we find it necessary to preempt state and local laws that effectively preclude the possession in vehicles or elsewhere of amateur service transceivers by amateur operators merely on the basis that the transceivers are capable of reception on public safety, special emergency, or other radio service frequencies, the reception of which is not prohibited by federal law.[37] We find that, under current conditions and given the types of equipment available in the market today, such laws prevent amateur operators from using their mobile stations to the full extent permitted under the Commission's Rules and thus are in clear conflict with federal objectives of facilitating and promoting the Amateur Radio Service. We recognize the state law enforcement interest present here, and we do not suggest that state regulation in this area that reasonably attempts to accommodate amateur communications is preempted.[38] This decision does not pertain to scanner laws narrowly tailored to the use of such radios, for example, for criminal ends such as to assist flight from law enforcement personnel. We will not, however, suggest the precise language that must be contained in state and local laws. We do find that state and local laws must not restrict the possession of amateur transceivers simply because they are capable of reception of public safety, special emergency or other radio service frequencies, the reception of which is not prohibited by federal law, and that a state or local permit scheme will not save from preemption an otherwise objectionable law.[39] Finally, we note, as stated by APCO in comments filed previously in this proceeding, that any public safety agency that desires to protect the confidentiality of its communications can do so through the use of technology such as scrambling or encryption.[40]

V. CONCLUSION

13. We hold that state and local laws that preclude the possession in vehicles or elsewhere of amateur radio service transceivers by amateur operators merely on the basis that the transceivers are capable of the reception of public safety, special emergency, or other radio service frequencies, the reception of which is not prohibited by federal law, are inconsistent with the federal objectives of facilitating and promoting the amateur radio service and, more fundamentally, with the federal interest in amateur operator's being able to transmit and receive on authorized amateur service frequencies. We therefore hold that such state and local laws are

preempted by federal law.

14. Accordingly, IT IS ORDERED that the request for a declaratory ruling filed by the ARRL IS GRANTED to the extent indicated herein and in all other respects IS DENIED.

FEDERAL COMMUNICATIONS COMMISSION

William F. Caton
Acting Secretary

APPENDIX

Comments or reply comments to the *Inquiry* were submitted by the following parties:

70 individual amateur operators, some of whom also operate GMRS equipment or use scanning receivers

2 individual General Mobile Radio Service (GMRS) operators

13 individual scanning receiver users

American Radio Relay League, Inc. (ARRL)

Associated Public-Safety Communications Officers, Inc. (APCO)

Bellcore Pioneers Amateur Radio Association

Big Spring/Howard County, Texas; Hal Boyd, Emergency Coordinator

C. Crane Company

Capital Cities/ABC, Inc.

City of Martinez, California; Gerald W. Boyd, Chief of Police

Communicatons Electronics, Inc.

County of Sussex, New Jersey; John Ouweleen, Emergency Management Coordinator

CO Communications, Inc.

Egyptian Radio Club, Inc.

Grove Enterprises, Inc.

Jessamine Amateur Radio Society

National Communications System (NCS), Department of Defense

Pasco County, Florida; Edith L. Sanders, Disaster Preparedness Coordinator

Personal Radio Steering Group, Inc. (PRSG)

Radio Communications Monitoring Association (RCMA)

Riverside County R.E.A.C.T.

Seminole County, Florida; Kenneth M. Roberts, Emergency Management Coordinator

State of Michigan, Department of State Police; David H. Held, Director, Communications Section

Tandy Corporation

Footnotes

1 The American Radio Relay League, Inc., Request for Declaratory Ruling Concerning the Possession of Radio Receivers Capable of Reception of Police or Other Public Safety Communications (November 13, 1989) (ARRL motion).

2 Transceivers are radio equipment capable of both transmission and reception.

3 Public Notice, 5 FCC Rcd 1981 (1990). 55 Fed. Reg. 10805 (March 23, 1990). Comments were due by May 16, 1990, and reply comments by May 31, 1990.

4 6 FCC Rcd 1305 (1991)(*Inquiry*).

5 The scanner laws appear to be aimed at promoting the health, safety, and general welfare of the citizenry.

6 *See generally* ARRL motion (citing N.J. Stat. Ann. §2A:127—4 (West 1985)(noting that a person is guilty of a misdemeanor for possessing or installing a short-wave radio in an automobile capable of receiving, *inter alia*, frequencies assigned for police use unless a permit has been issued therefor by the chief of the county or municipal police wherein such person resides) and Ky. Rev. Stat. Ann. §432.570 (Michie/Bobbs-Merrill 1985) (noting that any person who possesses a mobile short-wave radio capable of receiving frequencies assigned for police use is guilty of a misdemeanor , except that certain users such as radio and television stations, sellers of the "scanner" radios, disaster and emergency personnel, and those using the weather radio service of the National Oceanic and Atmospheric Administration are exempt, while amateur radio licensees are not exempt).

7 *See generally* ARRL motion (regarding, *inter alia*, a Kansas City, Missouri, scanner law). *See also* note 24, *infra*.

8 *Id*.

9 N.J. Stat. Ann. §2C:33-22 (West 1992).

10 Ky. Rev. Stat. Ann. §432.570(4)(c) (Baldwin 1992).

11 *See* note 22, *infra*.

12 ARRL Motion at 1, 3 and 5, "Most commercial amateur radio VHF and UHF transceivers...are incidentally capable of reception (but not transmission) on frequencies additional to those allocated to the Amateur Radio Service. These frequencies are adjacent to amateur allocations. This is true even though the equipment is primarily designed for amateur bands, and results from the intentional effort to insure proper operation of the transceiver throughout the entire amateur band in question." *Id*, at 3.

13 *Id*. at 12.

14 For example, amateur radio licensees were widely recognized as serving a vital role in providing communications from devastated areas of South Florida during Hurricane Andrew and its aftermath in 1992.

15 *See generally* House Comm. on the Judiciary, Electronic Communications Privacy Act of 1986, H.R. Rep. No. 647, 99th Cong., 2d Sess, 42.

16 ARRL motion at 2 and 12. As of February 28, 1993, the Commission's licensing database indicates that there are 598,656 amateur stations in the United States and its territories and possessions.

17 *Id*. at 11.

18 A list of commenters is provided in the Appendix. Further, we have accepted a comment from Communications Electronics, Inc., which was filed one day late. *See generally* 47 C.F.R. §1.46(b). We also have considered 45 comments filed previously in this proceeding. *See Inquiry*, 6 FCC Rcd at 1306-1308 (noting that all of the filed comments support the ARRL motion). In addition, we received comments from scanner (receive-only equipment) users, who are not federal licensees and whose interests have not been at issue in the proceeding.

19 Comment of State of Michigan, Department of State Police, at 2-3 (June 3, 1991). *But see* Reply Comments of Personal Radio Steering Group of Ann Arbor, Michigan (July 8, 1991)(noting that ARRL has not requested the preemption of state and local laws that proscribe unlawful actions taken by amateur licensees).

20 Comment of State of Michigan, *supra*, at 2-3. *But see* paragraph 12, n. 40, *infra* (noting the comments supporting preemption filed previously in this proceeding by the Associated Public Safety Communications Officers (APCO)).

21 *See, e.g.,* Comment of John F. Fuhrman at 4 (April 29, 1991), Comment of Joseph Reymann at 9, 14 (May 24, 1991), and Comment of Mark D. Tavaglini at 3 (July 5, 1991).

22 *See, e.g.,* Comment of ARRL at 12 & n.6, 14 (June 7, 1991); Comment of Association of North American Radio Clubs at 5 (April 30, 1990): Comment of Radio Communications Monitoring Association at 5 (June 6, 1991). With respect to scanner laws at the local level, ARRL has notes that it is difficult to determine the precise number of such ordinances. *See* Comment of ARRL at 12 (June 7, 1991); *See also* Letter from ARRL to the Chief, Private Radio Bureau, Federal Communications Commission, Washington, D.C. (May 26, 1993)(noting local scanner laws in effect in Newton and Overland Park,

Kansas, Jersey City, New Jersey, and Kansas City, Missouri).

[23] Comment of National Communications System at 2-4 (June 7, 1991).

[24] Capital Cities Cable, Inc. v. Crisp, 467 U.S. 691, 699-705 (1984) (*quoting* Rice v. Santa Fe Elevator Corp., 331 U.S. 218, 230 (1947)).

[25] Fidelity Fed. Savings & Loan Ass'n v. de la Cuesta, 458 U.S. 141, 153 (1982) (*quoting* Florida Lime & Avocado Growers, Inc. v. Paul, 373 U.S. 132, 142-43 (1963); Hines v. Davidowitz, 312 U.S. 52, 67 (1941)); *see* Capital Cities Cable, Inc. v. Crisp, 467 U.S. at 705-09.

[26] Fidelity Fed. Savings & Loan Ass'n v. de la Cuesta, 458 U.S. at 153.

[27] *Federal Preemption of State and Local Regulations Pertaining to Amateur Radio Facilities,* 101 FCC 2d 952, 959-60 (1985) (concerning amateur radio antenna restrictions) (*Amateur Preemption Order*), See 47 C.F.R. §97.1. *See also* Note. Federal Preemption of Amateur Radio Antenna Height Regulation: Should the Sky Be the Limit? 9 Cardozo L. Rev. 1501, 1517-19 (1988), Note, Local Regulation of Amateur Radio Antennae and the Doctrine of Federal Preemption: The Reaches of Federalism, 9 Pac. L.J. 1041, 1055-60 (1978).

[28] 47 C.F.R. §97.1(b).

[29] 47 U.S.C. §157(a).

[30] SENSE OF CONGRESS

Sec. 10

(a) The Congress finds that —

(1) more than four hundred thirty-five thousand four hundred radio amateurs in the United States are licensed by the Federal Communications Commission upon examination in radio regulations, technical principles, and the international Morse code;

(2) by international treaty and the Federal Communications Commission regulation, the amateur is authorized to operate his or her station in a radio service of intercommunications and technical investigations solely with a personal aim and without pecuniary interest;

(3) among the basic purposes for the Amateur Radio Service is the provision of voluntary, noncommercial radio service, particularly emergency communications; and

(4) volunteer amateur radio emergency communications services have consistently and reliably been provided before, during, and after floods, tornadoes, forest fires, earthquakes, blizzards, train wrecks, chemical spills, and other disasters.

(b) It is the sense of Congress that —

(1) it strongly encourages and supports the Amateur Radio Service and its emergency communications efforts; and

(2) Government agencies shall take into account the valuable contributions made by amateur radio operators when considering actions affecting the Amateur Radio Service.

Federal Communications Commission Authorization Act of 1988. Pub. L. No. 100-594, 102 Stat. 3021, 3025 (November 3, 1988); *see also* Joint Explanatory Statement of the Committee of Conference on H.R. Conf. Rep. No. 386. 101st Cong., 1st Sess. 415, 433 (November 21, 1989), *reprinted in 1990* U.S. Code Cong. & Admin. News 3018, 3037 (amateur licensees exempted from new Commission-wide fees program because "[t]he Conferees recognize that amateur licensees do not operate for profit and can play an important public safety role in times of disaster or emergency"). Joint Explanatory Statement of the Committee of Conference on H.R. Conf. Rep. No. 765, 97th Cong., 2d Sess. 18-19 (August 19, 1982), *reprinted in* 1982 U.S. Code Cong. & Admin. News 2261, 2262-63.

[31] *Id.*

[32] *Reorganization and Deregulation of Part 97 of the Rules Governing the Amateur Radio Services, Notice of Proposed Rule Making,* 3 FCC Rcd 2076, 2077, (1988), final rules adopted in *Report and Order,* 4 FCC Rcd 4719 (1989), aff'd in *Memorandum Opinion and Order,* 5FCC Rcd 4614 (1990).

[33] The rules, however, do prohibit amateur service *transmissions* outside of the allocated amateur service bands. 47 C.F.R §97.307(b); Public Notice, Extended Coverage Transceivers in the Amateur Radio Service, mimeo no. 4114 (July 21, 1987) (noting that "[i]t is a violation of the Commission's regulations to...transmit on a frequency allocated to a licensed service without the appropriate Commission-issued station license.").

[34] *See* para. 5, n.16, *supra.*

[35] *Cf.* Capital Cities Cable, Inc. v. Crisp, 467 U.S. at 711 (state ban on alcoholic beverages commercials preempted where compliance by cable companies might result in deletion of out-of-state programming, thereby frustrating federal goal of promoting programming variety).

[36] *See Amateur Preemption Order,* 101 FCC 2d at 960 (ordinances that "operate to preclude amateur operations in their communities are in direct conflict with federal objectives and must be preempted").

[37] We note that federal law prohibits unauthorized reception on frequencies of certain radio services, *e.g.,* cellular radio. *See* Electronic Communications Privacy Act of 1986. §§101(a)(1), 101(a)(6), 101(c), 18 U.S.C. §§2510(1), 2510(10), 2510(16)(d), 2511(1). House Comm. on the Judiciary, Electronic Communications Privacy Act of 1986, H.R. Rep. No. 647, 99th Cong., 2d Sess. 31-33, 37.

[38] *See Amateur Preemption Order,* 101 FCC 2d at 960 (state and local regulations regarding amateur antennas based on health, safety or aesthetic considerations "must be crafted to accommodate reasonably amateur communications and to represent the minimum practicable regulation to accomplish the local authority legitimate purpose").

[39] The possibility that an affected licensee might obtain an additional authorization or permit to operate under the state or local law does not ameliorate the conflict, because the state or local issuing authority might choose to deny the amateur operator the permit, or charge a fee for the permit, or require the permit of even a non-resident.

[40] *See* Comments of APCO at 2-3 (May 16, 1990) (summarized in *Inquiry,* 6 FCC Rcd at 1306).

Appendix 12
FCC Field Offices and Online Addresses

Protection for FCC monitoring facilities:

If your station is located one mile or less from an FCC Field Office with a monitoring facility (see list below), you must take steps to ensure you don't harmfully interfere with it. The FCC can put restrictions on your operating if you do interfere with their operations [97.13(b)]. Consult the facility's Engineer-in-Charge for advice.

Note: Field Office hours are 8:00AM-4:30PM local time.

*ALLEGAN, MICHIGAN OFFICE (AL)(monitor facility):   (616) 673-2063
PO Box 89   Fax (616) 673-2063
Allegan, MI 49010-9437

In the state of Indiana, the counties of: Allen, De Kalb, Elkhart, Fulton, Kosciusko, LaGrange, Marshall, Noble, St. Joseph, Steuben and Whitley.

In the state of Michigan, the counties of: Allegan, Antrim, Barry, Benzie, Berrien, Branch, Calhoun, Cass, Charlevoix, Clare, Eaton, Grand Traverse, Ionia, Isabella, Kalamazoo, Kalkaska, Kent, Lake, Leelanau, Manistee, Mason, Mecosta, Missaukee, Montcalm, Muskegon, Newaygo, Oceana, Osceola, Ottawa, St. Joseph, Van Buren and Wexford.

**ANCHORAGE, ALASKA OFFICE (AN) (monitor facility): (907) 243-2153
6721 West Raspberry Rd.   Fax (907) 243-2138
Anchorage, AK 99502-1896

All counties in the state of Alaska.

ATLANTA, GEORGIA OFFICE (AT):   (404) 279-4621
Suite 320, Koger Center-Gwinnett   Fax (404) 279-4633
3575 Koger Blvd
Duluth, GA 30136-4958

All counties in the states of Alabama, Georgia, South Carolina and Tennessee.

In the state of Florida, the counties of: Escambia and Santa Rosa.

*BELFAST, MAINE OFFICE (BE) (monitor facility):   (207) 338-4088
PO Box 470   Fax (207) 338-6403
Belfast, ME 04915-0470

In the state of Maine, the counties of: Androscoggin, Aroostook, Franklin, Hancock, Kennebec, Lincoln, Oxford, Penobscot, Piscataquis, Sagadahoc, Somerset, Waldo and Washington.

In the state of New Hampshire, the county of Coos.

BOSTON, MASSACHUSETTS OFFICE: (BS)   (617) 770-4023
NFPA Bldg.   Fax (617) 770-2408
1 Batterymarch Park
Quincy, MA 02169-7496

All counties in the states of: Connecticut, Massachusetts, Rhode Island and Vermont.

In the state of Maine, the counties of: York and Cumberland.

In the state of New Hampshire, the counties of: Belknap, Cheshire, Carroll, Grafton, Hillsborough, Merrimack, Rockingham, Strafford and Sullivan.

**BUFFALO, NEW YORK OFFICE: (BF)   (716) 551-3838
1307 Federal Building   Fax (716) 551-3817
111 West Huron St.
Buffalo, NY 14202-2398

In the state of New York, the counties of: Allegany, Broome, Cattaraugus, Cayuga, Chautauqua, Chemung, Chenango, Clinton, Cortland, Erie, Essex, Franklin, Fulton, Genesee, Hamilton, Herkimer, Jefferson, Lewis, Livingston, Madison, Monroe, Montgomery, Niagara, Oneida, Onondaga, Ontario, Orleans, Oswego, Otsego, St. Lawrence, Saratoga, Schoharie, Schuyler, Steuben, Tioga, Tomkins, Warren, Washington, Wayne, Wyoming and Yates.

CHICAGO, ILLINOIS OFFICE (CG):   (708) 298-5401
Park Ridge Office Center, Room 306   Fax (708) 298-5171
1550 Northwest Hwy
Park Ridge, IL 60068-1460

All counties in the states of: Illinois, Indiana (except those listed for the Allegan office) and Kentucky (except those counties listed for the Detroit office).

In the state of Wisconsin, the counties of: Brown, Calumet, Columbia, Crawford, Dane, Dodge, Door, Fond du Lac, Grant, Green, Iowa, Jefferson, Kenosha, Kewaunee, Lafayette, Manitowoc, Milwaukee, Outagamie, Ozaukee, Racine, Richland, Rock, Sauk, Sheboygan, Walworth, Washington, Waukesha and Winnebago. All others are covered by the St. Paul office.

COLUMBIA OPERATIONS CENTER
COLUMBIA, MD (monitor facility)   (410) 725-3474
PO Box 250   (410) 206-2896
Columbia, MD 21045-9998

In the state of Delaware, the counties of: Kent, Sussex and New Castle (below C & D Canal).

The District of Columbia.

All counties in the states of Maryland and West Virginia.

In the state of Virginia, the counties of: Arlington, Fairfax, Loudoun and Prince William. All others covered by Norfolk.

DALLAS, TEXAS OFFICE (DL):   (214) 235-3369
Suite 1170
9330 LBJ Freeway
Dallas, TX 75243-3429

All counties in the states of Oklahoma and Texas (except those listed for the Houston and Kingsville offices).

DENVER, COLORADO OFFICE: (DV)   (303) 969-6497
Suite 860   Fax (303) 969-6556
165 South Union Blvd
Lakewood, CO 80228-2213

All counties in the states of Colorado, New Mexico and Wyoming.

In the state of South Dakota, the counties of: Bennett, Butte, Corson, Custer, Dewey, Fall River, Gregory, Haakon, Harding, Jackson, Jones, Lawrence, Lyman, Meade, Mellette, Pennington, Perkins, Shannon, Stanley, Todd, Tripp and Ziebach.

All others are covered by the St. Paul office.

DETROIT, MICHIGAN OFFICE: (DT)   (810) 471-5605
24897 Hathaway St.   Fax (810) 471-0052
Farmington Hills, MI 48335-1552

In the state of Kentucky, the counties of: Bath, Bell, Boone, Bourbon, Boyd, Bracken, Breathitt, Campbell, Carter, Clark, Clay, Elliott, Estill, Fayette, Fleming, Floyd, Franklin, Gallatin, Garrard, Grant, Greenup, Harlan, Harrison, Jackson, Jessamine, Johnson, Kenton, Knox, Knott, Laurel, Lawrence, Lee, Leslie, Letcher, Lewis, Lincoln, Madison, Magoffin, Martin, Mason, McCreary, Menifee, Montgomery, Morgan, Nicholas, Owen, Ownsley, Pendleton, Perry, Pike, Powell, Pulaski, Robertson, Rockcastle, Rowan, Scott, Wayne, Whitley, Wolfe and Woodford.

All others are covered by the Chicago office.

All counties in the states of Michigan (except those listed for the Allegan and St. Paul offices) and Ohio.

*DOUGLAS, ARIZONA OFFICE (DS) (monitor facility):   (520) 364-8414
PO Box 6   Fax (520) 364-8414
Douglas, AZ 85608-0006

All counties in the state of Arizona (except La Paz and Yuma which are covered by the San Diego office).

In the state of Utah, the counties of: Emery, Garfield, Grand, Kane, Piute, San Juan, Sevier and Wayne.

*FERNDALE, WASHINGTON OFFICE (FE) (monitor facility): (360) 354-4892
1330 Loomis Trail Rd.   Fax (360) 354-4892
Custer, WA 98240-9303

In the state of Washington, the counties of: San Juan, Skagit and Whatcom.

*GRAND ISLAND, NEBRASKA OFFICE (GI) (monitor facility): (308) 381-4721
PO Box 1588   Fax (308) 381-4757
Grand Island, NE 68802-1588

All counties in the state of Nebraska.

**HONOLULU, HAWAII OFFICE (HL):   (808) 677-3318
PO Box 1030   Fax (808) 671-3352
Waipahu, HI 96797-1030

All counties in the state of Hawaii as well as the territories of: American Samoa, Guam, the Mariana Islands, Midway Island, Swains Island, Wake Island and all other Pacific Trust Territories.

**HOUSTON, TEXAS OFFICE (HU):   (713) 861-6200
1225 N. Loop West, Room 900   Fax (713) 861-0476
Houston, TX 77008-1775

In the state of Texas, the counties of: Angelina, Austin, Bastrop, Bexar, Blanco, Brazoria, Brazos, Burleson, Caldwell, Chambers, Colorado, Comal, De Witt, Fayette, Fort Bend, Galveston, Gillespie, Gonzales, Grimes, Guadalupe, Hardin, Harris, Hays, Jackson, Jasper, Jefferson, Kendall, Kerr, Lavaca, Lee, Liberty, Madison, Matagorda, Montgomery, Nacogdoches, Newton, Orange, Polk, Sabine, San Augustine, San Jacinto, Travis, Trinity, Tyler, Victoria, Walker, Waller, Washington, Wharton and Williamson.

KANSAS CITY, MISSOURI OFFICE (KC):   (816) 353-3773
8800 E. 63rd St. Room 320   Fax (816) 353-0611
Kansas City, MO 64133-4895

All counties in the states of Iowa, Kansas and Missouri.

*KINGSVILLE, TEXAS OFFICE (KI) (monitor facility):   (512) 592-2531
PO Box 632   Fax (512) 595-1938
Kingsville, TX 78364-0632

In the state of Texas, the counties of: Aransas, Atascosa, Bandera, Bee, Brooks, Calhoun, Cameron, Dimmit, Duval, Edwards, Frio, Goliad, Hidalgo, Jim Hogg, Jim Wells, Karnes, Kenedy, Kinney, Kleberg, La Salle, Live Oak, Maverick, McMullen, Medina, Nueces, Real, Refugio, San Patricio, Starr, Uvalde, Val Verde, Webb, Willacy, Wilson, Zapata and Zavala.

*LIVERMORE, CALIFORNIA OFFICE (LV) (monitor facility): (510) 447-3614
PO Box 311   Fax (510) 447-3622
Livermore, CA 94551-0311

In the state of California, the counties of: Alpine, Amador, Calaveras, Inyo, Mono, San Joaquin, Stanislaus and Tuolumne.

All counties in the states of Nevada and Utah (except those listed for the Douglas office).

LOS ANGELES, CALIFORNIA OFFICE (LA):   (310) 809-2096
Cerritos Corporate Tower   Fax (310) 865-0598
18000 Studebaker Rd, Room 660
Cerritos, CA 90701-3684

In the state of California, the counties of: Kern, Los Angeles, Orange, San Bernardino, San Luis Obispo, Santa Barbara and Ventura.

**MIAMI, FLORIDA OFFICE (MA): (305) 526-7420
Rochester Building, Room 310 Fax (305) 593-0399
8390 N.W. 53rd St.
Miami, FL 33166-4668

In the state of Florida, the counties of: Broward, Collier, Dade, Hendry, Lee, Monroe and Palm Beach. All others are covered by the Atlanta, Tampa or Vero Beach offices.

NEW ORLEANS, LOUISIANA OFFICE (OR): (504) 589-2095
800 West Commerce Road Fax (504) 733-0913
Room 505
New Orleans, LA 70123-3333

All counties in the states of Arkansas, Louisiana and Mississippi.

NEW YORK OFFICE (NY): (212) 620-3437
201 Varick St. Fax (212) 620-3718
New York, NY 10014-4870

In the state of New Jersey, the counties of: Bergen, Essex, Hudson, Hunterdon, Mercer, Middlesex, Monmouth, Morris, Passaic, Somerset, Sussex, Union and Warren. All others covered by the Philadelphia office.

In the state of New York, the counties of: Albany, Bronx, Columbia, Delaware, Dutchess, Greene, Kings, Nassau, New York, Orange, Putnam, Queens, Rensselaer, Richmond, Rockland, Schenectady, Suffolk, Sullivan, Ulster and Westchester. All others are covered by the Buffalo office.

**NORFOLK, VIRGINIA OFFICE (NF): (804) 441-6472
1200 Communications Circle Fax (804) 441-6474
Virginia Beach, VA 23455-3725

All counties in the states of North Carolina and Virginia (except those listed for the Baltimore office).

PHILADEPHIA, PENNSYLVANIA OFFICE (PA): (215) 752-1324
One Oxford Valley Office Building, Room 404 Fax (215) 752-2363
2300 East Lincoln Hwy.
Langhorne, PA 19047-1859

In the state of Delaware, the counties of: New Castle (above C & D Canal).

In the state of New Jersey, the counties of: Atlantic, Burlington, Camden, Cape May, Cumberland, Gloucester, Ocean and Salem. All others are covered by the New York office.

All counties in the state of Pennsylvania.

**PORTLAND, OREGON OFFICE (PO):   (503) 326-4114
1782 Federal Building   Fax (503) 326-7841
1220 SW 3rd Ave.
Portland, OR 97204-2898

   All counties in the states of Idaho (except those covered by the Seattle office) and Oregon.

   In the state of Washington, the counties of: Clark, Cowlitz, Klickitat, Skamania and Wahkiakum.

*POWDER SPRINGS, GEORGIA OFFICE (PS) (monitor facility): (770) 943-5420
PO Box 85   Fax (770) 943-4794
Powder Springs, GA 30073-0085

   All counties in the states of Alabama, Florida, Georgia, North Carolina, South Carolina, Tennessee and Virginia on an as needed basis. Otherwise coordinated with the Atlanta office for non-monitoring efforts.

**ST. PAUL, MINNESOTA OFFICE (PL):   (612) 774-5180
Suite 31   Fax (612) 290-3710
2025 Sloan Place
Maplewood, MN 55117-2058

   In the state of Michigan, the counties of: Alger, Baraga, Delta, Dickinson, Gogebic, Houghton, Iron, Keweenaw, Marquette, Menominee, Ontonagon and Schoolcraft. All others are covered by either the Allegan or Detroit office.

   All counties in the states of Minnesota, North Dakota, South Dakota (except those covered by the Denver office) and Wisconsin (except those covered by the Chicago office).

SAN DIEGO, CALIFORNIA OFFICE (SD):   (619) 467-0549
Interstate Office Park   Fax (619) 557-7158
4542 Ruffner St., Room 420
San Diego, CA 92111-2216

   In the state of Arizona, the counties of: La Paz and Yuma. All others are covered by the Douglas office.

   In the state of California, the counties of: Imperial, Riverside and San Diego.

SAN FRANCISCO, CALIFORNIA OFFICE (SF):   (510) 732-9046
3777 Depot Rd., Room 420   Fax (510) 732-6015
Hayward, CA 94545-1914

In the state of California, the counties of: Alameda, Butte, Colusa, Contra
Costa, Del Norte, El Dorado, Fresno, Glenn, Humboldt, Kings, Lake, Lassen,
Madera, Marin, Mariposa, Mendocino, Merced, Modoc, Monterey, Napa,
Nevada, Placer, Plumas, Sacramento, San Benito, San Francisco, San Mateo,
Santa Clara, Santa Cruz, Shasta, Sierra, Siskiyou, Solano, Sonoma, Sutter,
Tehama, Trinity, Tulare, Yolo and Yuba.

**SAN JUAN, PUERTO RICO OFFICE (SJ):   (809) 766-5567
San Juan Field Office   Fax (809) 766-5008
US Federal Building, Room 747
150 Carlos Chardon Avenue
Hato Rey, PR 00918-1731

All counties in the territories of Puerto Rico and the Virgin Islands.

SEATTLE, WASHINGTON OFFICE (ST):   (206) 821-9037
11410 NE 122nd Way   Fax (206) 820-0126
Suite 312
Kirkland, WA 98034-6927

In the state of Idaho, the counties of: Benewah, Bonner, Boundary, Clearwater,
Idaho, Kootenai, Latah, Lewis, Nez Perce and Shoshone. All others are covered by the Portland office.

All counties in the states of Montana and Washington (except those covered by
the Portland and Ferndale offices).

TAMPA, FLORIDA OFFICE (TP):   (813) 348-1508
Room 1215   Fax (813) 228-2872
2203 N. Lois Ave
Tampa, FL 33607-2356

In the state of Florida, the counties of: Duval (with assistance from Vero
Beach) plus all counties not covered by the Miami or Vero Beach offices.

*VERO BEACH, FLORIDA OFFICE (VB) (monitor facility):   (407) 778-3755
255 154th Ave   Fax (407) 778-3566
Vero Beach, FL 32968-9041

In the state of Florida, the counties of: Brevard, Flagler, Indian River, Martin,
Okeechobee, Orange, Osceola, St. Lucie, Seminole and Volusia.

*   These field offices will be closed in 1996. The monitoring facilities at these offices
    will be automated and operated remotely by the FCC's Columbia, Maryland,
    Operations Center. Other duties will be transferred to the remaining field offices.
** These field offices will be closed in 1996, but two members of the technical
    staff will be retained in these cities as resident enforcement agents.

FCC Information Available Online

Internet Access: You can use any of these Internet tools:

FTP: **ftp.fcc.gov**, log in as anonymous and use your e-mail address as the password. Publications are in the **/pub** directory and succeeding subdirectories. Identify files of interest by downloading the index (found at the same level as **/pub**) and searching for key words.

GOPHER: **fcc.gov** or use any gopher to get to "all the gophers in the world" then "U.S." then "DC" then "FCC".

World Wide Web: **http://www.fcc.gov**

FCC Forms via the Internet: You can obtain FCC forms via the Internet from the following addresses. At present, Forms 610 and 610-V are the only Amateur Service forms available. Forms are .zip files and when expanded become PC Paintbrush (.pcx) files. A Windows pcx reader is available in the same directory. Macintosh users note: these files can be opened and used with the right software.

FTP: **ftp.fcc.gov/pub/Forms**

GOPHER: **fcc.gov** then select the Forms directory

World Wide Web: **www.fcc.gov/Forms/**

Dial-in Access: The Industry Analysis Division of the Common Carrier Bureau maintains a public electronic bulletin board service (BBS), FCC-State Link, which provides information relating to the CCB. The BBS also holds general information about FCC actions (including the Daily Digest); it can be reached at (202) 418-0241.

Fax Access: The Daily Digest, News Releases, Speeches, selected Public Notices, and Job Announcements can be obtained from Fax-On-Demand at (202) 418-2830. Request the index(es) to find out the document number for documents you are interested in. Generally the document numbers are also listed on the fax copy of the Daily Digest.

Selected FCC forms and fact sheets can be obtained via Fax-On-Demand by calling (202) 418-0177 from the handset on your fax machine. Request the index to find out the document number for the forms you need. You must be calling from your fax machine to request a form. At press time, the only Amateur Service forms available by this service are the Form 610 and 610-V. If your fax output is on thermal paper, it must be photocopied before use; copies made on white paper are acceptable. If the original form is double-sided the copies should be double-sided. Forms must still be submitted by mail; fax or e-mail submissions will not be accepted.

Index

About The American Radio Relay League

The seed for Amateur Radio was planted in the 1890s, when Guglielmo Marconi began his experiments in wireless telegraphy. Soon he was joined by dozens, then hundreds, of others who were enthusiastic about sending and receiving messages through the air—some with a commercial interest, but others solely out of a love for this new communications medium. The United States government began licensing Amateur Radio operators in 1912.

By 1914, there were thousands of Amateur Radio operators—hams—in the United States. Hiram Percy Maxim, a leading Hartford, Connecticut, inventor and industrialist saw the need for an organization to band together this fledgling group of radio experimenters. In May 1914 he founded the American Radio Relay League (ARRL) to meet that need.

Today ARRL, with more than 170,000 members, is the largest organization of radio amateurs in the United States. The League is a not-for-profit organization that:
- promotes interest in Amateur Radio communications and experimentation
- represents US radio amateurs in legislative matters, and
- maintains fraternalism and a high standard of conduct among Amateur Radio operators.

At League headquarters in the Hartford suburb of Newington, the staff helps serve the needs of members. ARRL is also International Secretariat for the International Amateur Radio Union, which is made up of similar societies in more than 100 countries around the world.

ARRL publishes the monthly journal *QST*, as well as newsletters and many publications covering all aspects of Amateur Radio. Its headquarters station, W1AW, transmits bulletins of interest to radio amateurs and Morse code practice sessions. The League also coordinates an extensive field organization, which includes volunteers who provide technical information for radio amateurs and public-service activities. ARRL also represents US amateurs with the Federal Communications Commission and other government agencies in the US and abroad.

Membership in ARRL means much more than receiving *QST* each month. In addition to the services already described, ARRL offers membership services on a personal level, such as the ARRL Volunteer Examiner Coordinator Program and a QSL bureau.

Full ARRL membership (available only to licensed radio amateurs) gives you a voice in how the affairs of the organization are governed. League policy is set by a Board of Directors (one from each of 15 Divisions). Each year, half of the ARRL Board of Directors stands for election by the full members they represent. The day-to-day operation of ARRL HQ is managed by an Executive Vice President and a Chief Financial Officer.

No matter what aspect of Amateur Radio attracts you, ARRL membership is relevant and important. There would be no Amateur Radio as we know it today were it not for the ARRL. We would be happy to welcome you as a member! (An Amateur Radio license is not required for Associate Membership.) For more information about ARRL and answers to any questions you may have about Amateur Radio, write or call:

ARRL Educational Activities Dept
225 Main Street
Newington CT 06111-1494
(860) 594-0200
Prospective new amateurs call:
800-32-NEW HAM (800-326-3942)

FEEDBACK

Please use this form to give us your comments on this book and what you'd like to see in future editions, or e-mail us at **pubsfdbk@arrl.org** (publications feedback).

Where did you purchase this book?

☐ From ARRL directly ☐ From an ARRL dealer

Is there a dealer who carries ARRL publications within:

☐ 5 miles ☐ 15 miles ☐ 30 miles of your location? ☐ Not sure.

License class:

☐ Novice ☐ Technician ☐ Technician with HF privileges
☐ General ☐ Advanced ☐ Extra

Name	ARRL member? ☐ Yes ☐ No
_____	Call sign _____
Daytime Phone ( ) _____	Age _____
Address _____	
City, State/Province, ZIP/Postal Code_____	
If licensed, how long? _____	
Other hobbies _____	**For ARRL use only FCC RB**
_____	Edition 10 11 12 13 14 15 16 17
Occupation _____	Printing 2 3 4 5 6 7 8 9 10 11 12

From _____

_____

_____

EDITOR, FCC RULE BOOK
AMERICAN RADIO RELAY LEAGUE
225 MAIN ST
NEWINGTON CT 06111-1494

·· please fold and tape ····································